YOUR QUANTUM SELF

HOW MODERN PHYSICS AND ANCIENT WISDOM REVEAL WHO YOU ARE

ASHOK NARASIMHAN

For permissions, inquiries, or additional copies, contact:
Franklin Publishers

For my son Akshay: infinite in name and nature, eternally woven into the fabric of everything.

EARLY READER REVIEWS

In *Your Quantum Self* Ashok Narasimhan undertakes a cosmic journey of revelation that is based in modern physics and extends beyond particles, the known and unknown, to the heights of reality. Spacetime, quantum, the stars, are interwoven into the tapestry of the cosmos, leading back to the self through Vedanta and Shaivism and the ancient wisdom that these great systems of thought address. This is a true awakening and a roadmap that may lead us beyond the forest into the mystery that the book is seeking to understand. The journey leads to the ultimate mystery of what Consciousness truly is and opens the path to the center of one's being.

Overall, an exquisite book that connects the seemingly disparate threads of quantum physics and ancient wisdom to weave the fabric of reality that sends us on the most important journey of all: self-discovery.

– Menas Kafatos, PhD, Professor, co-author of New York Times bestseller *You Are the Universe*, Author of *Science, Reality & Everyday Life*

Modern physics, defined on the one hand by Einstein's space-time mass-energy continuum and on the other by quantum uncertainty and nonlocality of the Copenhagen interpretation, confronts us with a paradoxical world which our common conception has not been able to grasp in terms of a sensible representation.

In this remarkable book, Ashok Narasimhan puts forward the view that such a conceptual reimaging of reality is the need of our times and shows us the way to achieve this by bringing its phenomena into engagement with the profound realizations of ancient Indian wisdom teachers. The collision of two enigmatic systems, one contemporary and one ancient, results in brilliant flashes of transcendental intuition which illuminate both and transform our experience of reality for good.

- Debashish Banerji, PhD, Professor and Author, including *Time-Steps of the Cosmic Horse: The Contemplative Philosophy of the Great Forest (Brihadaranyaka) Upanishad*

A rare combination, part scientific narrative, part illumination, that finds ancient Indian philosophy speaking with uncanny clarity to modern physics. With a journalist's ear and a philosopher's inquisitiveness, the book shows how the metaphysics of Advaita Vedanta, composed millennia ago, anticipated the questions now animating the classical and quantum realms. The result is not mysticism by analogy, but a lucid parallelism that feels both intellectually rigorous and quietly revelatory.

- Isabel Sadurni, Film and Television Producer and Writer

In this fascinating book, Ashok Narasimhan takes us on a extraordinary voyage, melding memories, modern physics and ancient metaphysics to elucidate for curious minds the nature of reality, consciousness, and of the indivisibility of the individual self and the universe. In doing so, he also helpfully enables us to navigate the complex terrain between quantum science and transcendent philosophies of being, and to tackle the awesome symmetries of the self and the cosmos.

—Pillarisetti Sudhir, student of history and science, and former editor of *Perspectives on History*, the newsmagazine of the American Historical Association

Engaging, insightful, and full of practical wisdom, Ashok Narasimhan brings a fresh, integrative perspective that bridges quantum physics with ancient Vedic meditative, analytical enquiry. In this scenario, the quantum self is a logical extension and the only possible explanation of our oneness with the cosmos as well as one another. Written as a personal journey of self-discovery, this transformative guide is logical and so emotionally satisfying that it leads to an 'aha' moment wondering why we have ever thought anything different.

A masterful storytelling that makes quantum physics and Vedic analytical enquiry feel as intuitive as breathing. A must read for the curious mind.

- Neerja Raman, Author of *The Chemistry of Belonging: Stories of Inheritance and Upbringing* and *The Practice and Philosophy of Decision Making: A Seven Step Spiritual Guide*

CONTENTS

FOREWORD

We live in extraordinary times. Not just because of what science is discovering—but because of what those discoveries are beginning to reveal about us. About reality. About what lies beneath the surface.

This book is for the curious. For the seekers. For the quietly thoughtful, the deeply questioning, the intellectually open-hearted. It's for those who look up at the stars or into their own reflection—and wonder: What is all this? What am I?

You'll find no academic rigor-mortis here. I've done that in my published scientific papers. This is not one of them. This book joins a growing movement of work that attempts to bridge the worlds of science and inner experience. But my aim is not to preach a synthesis or offer platitudes. Instead, I invite you into something more intimate—a discovery of your true nature as a quantum self.

You don't need a Ph.D. to engage with these ideas. All you need is a mind that's curious and a heart that's willing to listen. If you find yourself somewhere between the rational and the reflective, if you're drawn to science but also aware there's more to life than logic alone—then this book was written for you. You might be thirty-five, urban or suburban, educated but still searching. Or maybe you're sixty-five, having seen the pendulum swing and now seeking something beyond the edges of mainstream understanding.

This book is for the spiritually conscious, the intellectually hungry, and the wellness-minded explorer. Those who suspect, perhaps quietly, that the self isn't just a mind in a body but a node in a larger matrix of being.

My own journey took me from the structured beauty of physics to the luminous truths hidden in ancient Indian wisdom. Along the way, I encountered something both startling and affirming: that the deepest insights of the ancient Indian philosophical traditions of Advaita Vedanta and Kashmir Shaivism mirror, in poetic form, what modern quantum physics is beginning to unveil. That we are not separate from reality, but expressions of it.

In both traditions—one contemplative, the other empirical—the world is not made of isolated objects, but of vibrational potential and deeply interwoven awareness. This convergence is not a metaphor. It is a possibility. One that changes everything.

So here's my invitation: walk with me through these revelations. We'll travel from light to meaning, from matter to consciousness. We'll explore ancient insights refracted through the lens of modern science. And perhaps—just perhaps—we'll catch a glimpse of what has been there all along.

Welcome to the adventure of a lifetime—the journey home to your quantum self.

—Ashok Narasimhan

PROLOGUE

"The universe is not composed of physical things, but of things that exist in a strange realm, beyond space and time. These things are not material, but immaterial; they are forms, structures, and patterns"

— Werner Heisenberg, Nobel Prize in Physics, 1932

The Night Everything Changed

I remember the exact moment my understanding of reality shattered. It wasn't in a laboratory or lecture hall, but standing beneath a canopy of stars on a moonless night in the mountains of Colorado. The Milky Way stretched across the heavens like a river of light, and I found myself asking a question that would consume the next decades of my life: If everything I can see—every star, every galaxy, every particle of light traveling across billions of years to reach my eyes—follows the elegant laws of physics I've studied, why does the quantum world seem to operate by completely different rules?

To understand why this question haunted me, you need to know something remarkable about the universe we inhabit. The world we experience—the world of falling apples, orbiting planets, and light traveling predictably through space—operates according to rules that feel almost intuitive once you learn them. Objects have definite positions. Causes precede effects. Things exist in one place at a time. This is the classical realm, the spacetime we navigate every day.

But venture down to the scale of atoms and their constituent particles, and you enter a domain where everything you think you know about reality dissolves. In this quantum realm, particles seem to exist in multiple states simultaneously until observed. They appear to communicate instantaneously across vast distances. The very act of measuring something changes what you're measuring. It's as if the universe operates according to two entirely different sets of instructions—one for the large and familiar, another for the impossibly small—and nobody has been able to explain why these two realities should be so radically different, or how they connect.

That question, so simple to ask and so impossible to answer, launched me on a journey I never expected. It would lead me through the strangest territories of modern physics, into laboratories where particles behave as if they can be in two places at once, through thought experiments that had tormented Einstein himself, and—most unexpectedly—into the contemplative traditions of ancient India, where I found descriptions of reality that seemed to anticipate what our most sophisticated instruments were only now beginning to reveal.

What I discovered along the way didn't just satisfy my intellectual curiosity. It transformed my understanding of what it means to exist, to be conscious, to be human. And it began, of all places, in a forest.

The Forest That Changed How I See

The breakthrough came from an unexpected source: walking through an ancient forest and really seeing it for the first time.

Above me stretched a magnificent canopy—what ecologists call the overstory—where sunlight danced through leaves, birds nested in branches, and life unfolded according to familiar patterns I could observe and measure. This visible realm followed predictable rules: photosynthesis converted sunlight to energy, water flowed downward, cause led to effect in a reassuring sequence.

But beneath my feet lay another world entirely—the understory of tangled roots, fungal networks, and hidden communication systems that sustained the forest from below. Here, trees "talked" to each other through chemical signals, sharing nutrients across impossible distances. Mother trees nourished their offspring through underground networks spanning entire forests. Mycorrhizal fungi formed a kind of "wood wide web" where information and resources flowed in patterns that defied everything I thought I knew about individual organisms.

Standing there, with my feet on the hidden understory and my eyes on the visible canopy, a thought struck me with the force of revelation: What if reality itself worked exactly like this forest?

What if our familiar world of space and time—the realm where planets orbit stars and galaxies dance their cosmic ballet—was just the overstory of existence? And what if beneath it lay a hidden understory operating by entirely different principles?

The more I explored this possibility, the more I began to suspect that the quantum realm wasn't just scientifically interesting—it was the key to understanding everything we think we know about reality. And perhaps about ourselves.

The Mysteries That Haunted Me

This insight sent me diving deep into the strangest phenomena in physics, and what I discovered defied everything I thought possible.

Consider light. We swim in it, see by it, measure time by it. Yet light exists in a state that seems to mock our understanding. From a photon's perspective, Einstein's equations tell us something impossible: space vanishes and time stops completely. A photon traveling from a distant star experiences no journey at all—departure and arrival are the same instant. Yet we observe light every day in our familiar three-dimensional world, moving at a constant speed, taking time to cross a room. How can something exist both within time and utterly outside it?

Then there was the double-slit experiment, which has been called the most beautiful experiment in physics—and the most disturbing. Fire electrons one at a time at a barrier with two slits, and they create interference patterns as if each electron passes through both slits simultaneously. But try to determine which slit each electron actually uses, and the pattern vanishes. The electron "chooses" a single path. It's as if reality itself responds to whether or not we're paying attention.

And then there was entanglement—the phenomenon that nearly broke my understanding entirely. When two particles become entangled, they remain connected regardless of distance. Measure one particle, and you instantly know the state of its partner, even if it's on the other side of the universe. Einstein called this "spooky action at a distance" and spent years trying to prove it couldn't be real. He failed. Experiments have confirmed that entanglement is absolutely real, that particles can share a connection that transcends space itself.

These weren't just laboratory curiosities. They were cracks in the foundation of everything I thought I understood about how the universe works. And as I stared into those cracks, I began to glimpse something vast moving beneath the surface of the familiar world.

The Ancient Map I Never Expected to Find

As I grappled with these impossibilities, I found myself drawn—almost against my will—to ancient philosophical traditions that seemed to describe exactly what modern physics was discovering. In the 3,000-year-old texts of Advaita Vedanta, I encountered descriptions of reality as having both manifest and unmanifest aspects. In Kashmir Shaivism, I found an intricate framework describing how the formless gives rise to form, how consciousness and energy dance together to create the world we experience.

At first, I resisted. I had seen too many instances where people indulged in a kind of intellectual solipsism, claiming that all of modern science was already known to the ancients—usually through vague pattern-matching and wishful thinking. "The atom was described in the Vedas!" "Quantum physics proves Eastern mysticism!" These claims almost always collapsed under scrutiny, revealing more about the claimant's desire for validation than about any genuine correspondence between ancient insight and modern discovery.

So I approached these texts with deep skepticism, expecting to find the same loose analogies dressed up as profound connections. What I found instead stopped me cold. The parallels were too precise to dismiss. These weren't vague spiritual intuitions that happened to sound poetic. They were rigorous descriptions of reality's structure—developed through systematic contemplative investigation rather than particle accelerators, but arriving at conclusions that mapped onto the quantum mysteries with uncanny precision.

I began to wonder: What if these ancient traditions and modern physics were two different expeditions that had explored the same territory from opposite directions? What if they had drawn maps of the same landscape using entirely different methods—

one through mathematics and experimentation, the other through systematic introspection—and found themselves describing the same underlying reality?

The possibility was intoxicating. And terrifying. Because if it was true, then everything I thought I knew about consciousness, about matter, about my own existence, was about to be called into question.

The Questions That Drive This Book

What you hold in your hands is the record of what I found when I followed these questions wherever they led. It is the story of six quantum mysteries that, when understood together, reveal something profound about the nature of reality—and about who we are.

How does light manage to exist both within time and beyond it? What does it mean that observation seems to affect reality at the most fundamental level? How can particles remain connected across vast distances, as if space itself is an illusion? What is the mysterious dark energy that fills the cosmos, and what does it tell us about the creative potential hidden in apparent emptiness? Is spacetime itself fundamental, or does it emerge from something deeper? And how does a single fertilized cell know how to become a human being, coordinating trillions of processes with a precision that seems to exceed what classical physics can explain?

Each of these mysteries, I discovered, is a doorway. And what waits on the other side is not just a new theory about the universe, but a transformed understanding of what you are.

The Forest Calls

The ancient forest where this journey began still whispers its secrets to anyone willing to listen. Its visible canopy and hidden root system continue their eternal dance, reminding us that reality may be far more layered, more intimately interconnected, and more alive than we ever dared imagine.

As we prepare to enter this deeper understanding together, I want to make you a promise: I will not ask you to accept anything on faith. Every insight in this book can be examined, questioned, tested against your own experience and understanding. The quantum mysteries are real, documented by rigorous experiments. The ancient traditions offer frameworks that can be explored through your own contemplation. And the synthesis I'm proposing stands or falls on whether it actually illuminates your experience of being alive.

But I will also give you a warning: This journey may change how you see everything. The questions we're about to explore don't just rearrange intellectual furniture—they open trapdoors into a deeper understanding of existence itself. Once you glimpse the hidden understory of reality, the overstory never looks quite the same.

The forest of reality beckons, with its visible wonders and hidden depths waiting to be explored. Are you ready to discover what lies beneath the surface of the world you thought you knew?

The journey begins now.

CHAPTER ONE

The Overstory and Understory of Reality

"Everything we call real is made up of things that cannot be regarded as real"

— Niels Bohr, Nobel Prize in Physics, 1922

Bohr's paradox haunted me for years. How could reality be built from unreality? How could the solid world emerge from ghostly quantum mists? The answer came not from any laboratory, but from walking through an ancient forest—and from wisdom whispered by sages three millennia ago.

The Forest's Secret

Stand in any woodland and you witness reality's greatest magic trick. Above spreads a magnificent canopy—the overstory where light dances through leaves, where birds nest and seasons paint their colors. This visible realm follows predictable patterns: photosynthesis converts sunlight to energy, water flows downward, cause leads to effect. Everything here can be touched, measured, predicted with mathematical precision.

Watch a maple leaf catch the morning light, and you're observing a process that follows Newton's laws of motion and thermodynamics. The chlorophyll molecules absorb specific wavelengths with mechanical precision. Water molecules rise through xylem tubes according to well-understood principles of capillary action and transpiration. When autumn arrives, the breakdown of chlorophyll reveals carotenoids and anthocyanins in a chemical symphony as predictable as clockwork.

The visible forest operates like a vast, complex machine where every component has its place and function. Songbirds time their migrations to celestial rhythms that have persisted for millions of years. Predator and prey populations oscillate in cycles that can be mapped mathematically. Even the seemingly chaotic fall of leaves follows aerodynamic principles that engineers use to design aircraft.

But beneath your feet lies another world entirely—the understory of tangled roots, fungal networks, and microbial communities. Here, trees communicate through chemical signals, sharing nutrients across impossible distances. A mother tree nourishes her offspring through underground networks that span entire forests. Mycorrhizal fungi form a "wood wide web" where information and resources flow in patterns that would mystify any surface observer.

This hidden realm operates by rules utterly foreign to the visible world above. A Douglas fir can recognize its own seedlings and send them more nutrients than it provides to strangers. When a tree is attacked by insects, it releases chemical warnings that alert other trees to begin producing defensive compounds before the threat arrives. Fungi can transport nitrogen from nitrogen-rich areas to nitrogen-poor ones, phosphorus in the opposite direction, and carbon to wherever it's most needed—all simultaneously, through the same network.

Most remarkably, this underground economy includes trade negotiations, seasonal contracts, and even what appears to be lending with interest. Trees that receive nutrients during their struggling early years will later "repay" the network when they become mature producers. Some fungal networks appear to manipulate their tree partners, encouraging them to produce more sugars in exchange for minerals, like underground brokers facilitating complex deals.

The two realms exist in constant conversation—the underground networks determining which trees flourish in the canopy, while the canopy's photosynthesis feeds the root world below. Cut one tree, and the underground web instantly begins rerouting resources to compensate. Remove a key fungal species, and the visible forest will show signs of stress within weeks.

Neither can exist without the other, yet most observers see only the visible forest and miss the hidden half of reality. We walk through a world of apparent individuals—separate trees competing for sunlight and space—while below our feet unfolds a vast cooperative intelligence that makes the internet look simple by comparison.

The revelation struck me like lightning: reality itself operates exactly like this forest.

Einstein's Cosmic Dance Floor

Our familiar world—what I call the SpacetimeFabric—is reality's overstory. This is Einstein's magnificent discovery: space and time woven into a single, flexible continuum where galaxies waltz in their ancient choreography and planets trace precise orbits. Here, your coffee steams at a predictable rate, light travels its measured path from distant stars, and GPS satellites know exactly where they are.

Einstein showed us that this fabric has its own rhythm, its own responsiveness to the mass and energy that moves through it. When his special relativity emerged from the ashes of the failed search for the luminiferous ether, it revealed spacetime as a dynamic, responsive medium rather than the fixed stage that Newton had imagined.

The Michelson-Morley experiment had wiped the cosmic blackboard clean with results so unexpected they shook the foundations of physics. Scientists had spent decades searching for the ether—that hypothetical medium that was supposed to carry light waves through space. If Earth was moving through this ether, there should be a detectable "ether wind" that would affect the speed of light in different directions.

Albert Michelson and Edward Morley designed an exquisitely sensitive interferometer to detect this wind. They split beams of light, sent them on perpendicular journeys, and recombined them to look for the telltale interference patterns that would reveal Earth's motion through the ether. The experiment was so precise it could detect differences in the speed of light smaller than one part in a billion.

The results were stunning: no interference pattern, no ether wind, no difference in light's speed in any direction. The ether—that supposed foundation of electromagnetic theory—simply didn't exist. The cosmic blackboard had been wiped clean, and physicists faced a crisis that would require a complete reconceptualization of space and time themselves.

Einstein, working as a patent clerk with nothing but radical intuition, proposed the impossible: light's speed is absolute, the same for all observers regardless of their motion. This wasn't just a minor correction to existing theory—it was a complete overthrow of our understanding of space and time.

This simple statement unleashed a revolution that would transform not just physics but our entire conception of reality. Move fast enough, and time itself stretches like taffy. Approach light speed, and space contracts while time slows to a crawl. Twin siblings can age at different rates, clocks can tick at different tempos—all because spacetime itself is alive, responsive, dynamic.

The implications were staggering and counterintuitive. Time, which had seemed absolute and universal, became relative and personal. Space, which had appeared fixed and unchanging, became malleable and dynamic. The very stage upon which cosmic drama unfolds was revealed to be an active participant in that drama.

Consider the GPS satellites circling Earth at 12,000 miles above the surface. These satellites experience time running slightly faster than clocks on Earth's surface—about 38 microseconds per day faster, due to both their speed (special relativistic effect) and their position in Earth's gravitational field (general relativistic effect). Without constant corrections for these effects, GPS systems would accumulate errors of several miles per day. The satellites must constantly adjust their timing to account for the fluid nature of spacetime itself.

But then Einstein's equations revealed something impossible.

Light's Impossible Secret

Light, the very agent that choreographs spacetime's dance, exists outside its own ballroom. From a photon's perspective, Einstein's mathematics tell us, space vanishes and time stops completely. A photon experiences no journey—it is simultaneously everywhere along its path, existing in an eternal now that spans the entire universe.

This creates a mind-bending paradox that challenged everything physicists thought they understood. Light sets the speed limit for spacetime, defines the very structure of Einstein's fabric, yet from its own reference frame, it doesn't exist "in" spacetime at all. For a photon traveling from the edge of the observable universe to your eye, billions of light-years collapse to zero distance, and billions of years compress to no time at all.

The mathematics are unambiguous but the implications are staggering. According to special relativity, as an object approaches the speed of light, its length contracts toward zero in the direction of motion, while time dilation causes its internal clocks to slow toward a complete stop. At exactly the speed of light—the speed that photons always travel—length contraction becomes complete and time dilation becomes infinite.

From a photon's perspective, if it could have one, the entire universe collapses to a single point and all of cosmic history occurs in a single instant. The photon that left a distant star billions of years ago and the moment it strikes your retina are the same event from the photon's frame of reference. Past, present, and future collapse into an eternal now.

How can the choreographer of spacetime's dance be exempt from its own rules? How can the entity that weaves space and time together exist beyond space and time? How can something be simultaneously the most fundamental aspect of our physical reality and completely outside the framework that defines physical reality?

This paradox went deeper than mere mathematical curiosity. Light carries energy and momentum, creates electromagnetic fields, and drives every chemical reaction that powers life. The photons illuminating these words, the electromagnetic radiation that carries signals through your nervous system, the solar energy that drives Earth's weather—all of it exists in this impossible state outside spacetime while simultaneously being the foundation of everything that happens within spacetime.

This was my first glimpse into reality's hidden understory.

If light—the foundation of everything we see—doesn't actually experience space and time, then where does it exist? And what does this tell us about the nature of reality itself?

The Ancient Key

The answer came from the most unexpected source: verses I'd studied years earlier from the Upanishads, sacred texts composed over three thousand years ago in India:

The Brhadaranyaka Upanishad (Verse 3.8.8), says:

"It moves, and It moves not; It is far, and It is near; It is within all this, and It is also outside all this."

The It in this verse refers to Brahman, the ultimate reality underlying all existence. Note that Brahman is not a god or a deity, and not to be confused with Brahmin, which is a caste.

When ancient sages spoke of Brahman—the ultimate reality underlying all existence—they described exactly what modern physics was discovering about light and the quantum realm. Something that transcends space and time yet permeates every particle of our world. Something that is both beyond the manifest universe and intimately woven through every atom.

At first glance, these ancient words seem like mystical poetry, riddled with contradictions. How can something move and not move? How can it be simultaneously far and near, within and without? But as I grappled with the photon's impossible existence, these "contradictions" suddenly revealed themselves as precise descriptions of a reality beyond our ordinary categories of thought.

"It moves, and it moves not; it is far and it is near". These words say that it transcends space and time. It is not bound by the limitations of our physical universe, for it is the source and sustainer of all that is. Brahman is the eternal, unchanging essence that permeates every corner of the cosmos, the immutable foundation upon which the fleeting dance of creation unfolds. And yet, even as it remains forever beyond the reach of our senses and the grasp of our intellect, Brahman is intimately interwoven into the very fabric of our existence.

"It is within all this, and It is also outside all this". This is the great paradox and the great wonder of Brahman: that it is both transcendent and immanent, both beyond and within. It is the ground of all being, the source from which all things arise and to which all things return.

And yet, it is not separate from the world of form and phenomena, but rather infuses every atom of creation with its divine presence. Just as the sun's rays illuminate the earth and give life to all that grows, so too does Brahman pervade and animate the universe, from the vastness of galaxies to the subtlest of subatomic particles.

The quantum realm, like the Vedic concept of "It," or Brahman, is a domain of pure potentiality. Just as "It", Brahman, contains within itself the potential for all things, the quantum realm holds the infinite possibilities for a particle's existence. A quantum particle, before it is observed, is not confined to a single, definite state, but rather exists as a wave function, a mathematical description of *all its potential states*. In this sense, the quantum realm is the source from which all manifest reality emerges, just as Brahman is the source of all creation.

Moreover, the quantum realm's influence extends far beyond individual measurement events. Even after a particle manifests definite properties in spacetime through observation, quantum effects continue to shape its behavior. While the wave function collapse determines specific outcomes at the moment of measurement, quantum mechanics governs how the particle evolves between measurements, creating the probabilistic patterns we observe in quantum systems.

In this way, the quantum realm remains interwoven with the fabric of spacetime, just as Brahman is described as being intimately connected with the manifest world.

The assertion that Brahman is "within all this, and also outside all this" beautifully captures the essence of the vision of what I call the Unified Conscious Matrix. The quantum threads woven in the substrate of reality, is indeed "within all this," underlying every particle and force in the universe. At the same time, it exists in a domain beyond our familiar spacetime, in a sense "outside all this." The SpacetimeFabric, on the other hand, is the manifest universe we inhabit, yet it too is part of something greater, woven into the fabric of the Unified Conscious Matrix.

The Upanishads weren't products of primitive thinking or religious wishful thinking. They emerged from a sophisticated philosophical tradition that had developed precise methods for investigating the nature of consciousness and reality. The sages who composed these texts were as rigorous in their investigations as any modern scientist, but they used the laboratory of consciousness itself rather than external instruments.

These ancient investigators had developed what they called "direct perception" (pratyaksha)—a method of inquiry that bypassed the limitations of sensory experience and conceptual thinking to apprehend reality as it actually is, rather than as it appears through the filters of perception and cognition. Through sustained practice of meditation and contemplative inquiry, they claimed to have discovered the fundamental nature of existence itself.

What they found was that the ultimate reality—which they called Brahman—existed beyond all the dualities that structure ordinary experience. It was neither existent nor non-existent in the conventional sense, neither moving nor still, neither one nor many. It was the ground from which all dualities emerge while remaining forever beyond them.

The sages weren't speaking metaphorically—they were mapping the architecture of existence itself. They understood that ultimate reality operates beyond the dualities that structure our everyday experience. Just as light transcends the spacetime framework it creates, Brahman exists beyond the polarities of movement and stillness, distance and proximity, inside and outside.

Entering the QuantumLoom

Following light's impossible trail led me into quantum mechanics—reality's understory that I call the QuantumLoom. Here, in a realm beyond space and time, particles exist as waves of pure possibility, spread across mathematical landscapes called Hilbert space. A quantum particle isn't *somewhere*—it's *everywhere-at-once* until the moment we observe it.

This quantum realm operates in the abstract domain of Hilbert space, a multidimensional mathematical construct where particles exist as probability waves rather than definite objects. Imagine a vast landscape where every point represents a possible state of a particle—its potential position, momentum, energy, or spin. In this strange domain, a photon doesn't occupy a single point but spreads as a ghostly presence across the entire space of possibilities.

The mathematics of quantum mechanics describe this ghostly realm with extraordinary precision. The *wave function*—the fundamental equation of quantum theory—specifies the probability amplitude for finding a particle in any given state. But these aren't ordinary probabilities like the chance of rolling dice. They're quantum amplitudes that create the strange phenomenon of quantum superposition where particles can exist in multiple states simultaneously.

Reality exists in these multiple states simultaneously until measurement collapses the wave of possibilities into a single, definite outcome. A radioactive atom exists in a superposition of having decayed and not having decayed. Schrödinger's famous cat, isolated in a box with a Geiger counter connected to a vial of poison, exists in a superposition of being alive and dead until someone opens the box to look.

But here's what puzzled me most: when we observe a quantum particle, we don't encounter it in its abstract Hilbert space domain. We measure it in our familiar spacetime, feeling its impact on our instruments, registering its energy on our detectors. The particle somehow bridges these two realms, existing simultaneously in the timeless quantum domain and the temporal classical world.

The mathematics describe this perfectly, but mathematics alone couldn't explain the deeper mystery: how do these two realms—the visible SpacetimeFabric and the hidden QuantumLoom—relate to each other?

The Vedantic Breakthrough

Again, ancient wisdom provided the key. The more I studied Advaita Vedanta, the more I realized these sages had intuited something that modern physics is only now rediscovering through experimental verification: apparent opposites are often complementary aspects of a deeper unity.

The Vedantic understanding of Brahman became my roadmap through quantum mysteries. When the ancient texts speak of Brahman as both transcendent and immanent—simultaneously beyond spacetime and permeating every atom—they're describing exactly what we observe in quantum mechanics. The quantum realm exists "outside" spacetime in the sense that it operates beyond spatial and temporal constraints, yet it's "within" spacetime as the foundation from which all classical phenomena emerge.

The Vedantic tradition had developed sophisticated methods for understanding the relationship between the manifest and unmanifest aspects of reality. They distinguished between Brahman as Nirguna (without attributes) and Saguna (with attributes), but insisted these weren't separate realities but complementary perspectives on one indivisible truth.

Nirguna Brahman was described as beyond all qualities, beyond space and time, beyond existence and non-existence in any conventional sense. It was pure consciousness itself—not consciousness *of* anything, but the very ground of awareness from which all possible experiences emerge. This formless, attributeless reality was the ultimate foundation of existence.

Saguna Brahman was the same reality appearing with qualities, manifesting as the world of forms and phenomena. This wasn't a different reality but the same Brahman expressing itself through the play of maya—the creative power that manifests multiplicity within unity, diversity within oneness.

This insight from Advaita Vedanta revolutionized my understanding of the measurement problem in quantum mechanics. The ancient sages taught that Brahman doesn't "become" the manifest world through some temporal process—the manifest and unmanifest are simultaneous expressions of a single reality. Similarly, quantum possibilities don't "turn into" classical actualities through a process in time. Rather, the quantum and classical aspects are complementary faces of one unified phenomenon.

The ocean and wave metaphor from Vedantic literature provided another crucial insight. Just as waves are not separate from the ocean but are temporary expressions of the ocean's movement, classical particles are not separate from the quantum field but are localized expressions of quantum possibility. The wave doesn't become the ocean, nor does the ocean become the wave—they are one reality expressing itself in different modes.

This understanding suggested that the apparent problem of wave function collapse—how quantum superpositions become classical definite states—was based on a false premise. There was no real collapse, no transition from one type of reality to another. There was only one reality expressing itself through complementary aspects that appeared different depending on the scale and mode of observation.

This Vedantic understanding suggested that the quantum realm and spacetime aren't separate domains that somehow interact, but rather like the forest's visible canopy and hidden root system—complementary aspects of a single, living reality. Just as Brahman is simultaneously transcendent (beyond spacetime) and immanent (pervading all of spacetime), reality's quantum and classical aspects aren't separate domains but twin expressions of an underlying wholeness.

The Photon's Revelation

The photon's impossible existence suddenly made perfect sense through this Vedantic lens. It doesn't travel between two separate realms—it reveals that these realms were never separate to begin with. The SpacetimeFabric and QuantumLoom are complementary aspects of a single, living reality.

When a photon manifests in our spacetime instruments, it's not crossing a boundary from quantum to classical. Instead, it's revealing different aspects of its unified nature depending on how we interact with it. This mirrors the Vedantic teaching that Brahman appears different to different observers not because Brahman changes, but because our modes of perception reveal different facets of the one reality.

The wave-particle duality that had puzzled physicists for decades suddenly made sense. Light wasn't sometimes a wave and sometimes a particle—it was always one unified phenomenon that appeared wavelike or particle-like depending on how we chose to observe it. The complementarity principle of quantum mechanics was simply a manifestation of the deeper truth that reality transcends all our conceptual categories while manifesting through them.

The mathematics of quantum mechanics perfectly describe this complementarity. Decoherence theory shows how quantum superpositions transform into classical definiteness without requiring any collapse mechanism. Quantum information theory reveals how classical information emerges from quantum correlation without any fundamental divide between quantum and classical realms.

The measurement problem—long considered one of the deepest puzzles in physics—dissolved when viewed through this Vedantic lens. What we call the collapse was only one unified reality expressing itself through different modes of interaction and observation.

The Unified Conscious Matrix

This insight transformed everything. What we call reality isn't a collection of separate objects floating in empty space—it's a unified field that I have called the Unified Conscious Matrix, where quantum potentiality and spacetime actuality dance together in eternal partnership.

Consider how this resolves physics' greatest puzzle. For decades, scientists have struggled to reconcile quantum mechanics with general relativity because they treated these as descriptions of separate worlds. Quantum mechanics describes the microscopic realm where uncertainty and probability reign, while general relativity describes the macroscopic world where spacetime curves and gravity flows. The two theories seem to speak different languages about different realities.

But when we recognize them as complementary aspects of a unified reality—like studying a forest's canopy and roots as parts of one ecosystem—the apparent contradictions dissolve. They're simultaneous expressions of a deeper unity, like two sides of a single coin or the complementary strands of DNA's double helix.

The Cosmic Atelier

Picture a cosmic atelier where reality itself is crafted. At one end stands the magnificent SpacetimeFabric, stretched across the universe like a grand tapestry where galaxies waltz and stars are born. This is the realm we can touch, see, and measure—the canvas of our everyday existence.

But this isn't a canvas waiting passively to be painted upon. The SpacetimeFabric is itself alive, responsive, dynamic. It curves in response to energy and momentum, creating the phenomenon we call gravity. It ripples with gravitational waves when massive objects accelerate. It expands as the universe grows, carrying galaxies along like leaves on a flowing river.

Turn your gaze to the other end of this celestial workshop, and you discover the QuantumLoom, a device so intricate and mysterious that it defies our macroscopic understanding. Here, quantum threads shimmer with possibility, existing in multiple states simultaneously. Photons and particles dance in an eternal, spaceless ballet, their movements weaving the very essence of reality.

The QuantumLoom operates in dimensions beyond the three spatial dimensions and one time dimension of our familiar experience. It weaves patterns in Hilbert space—abstract mathematical realms where probability amplitudes interfere and combine according to the strange logic of quantum mechanics. Every possible configuration of matter and energy, every potential outcome of every interaction, exists as a thread in this cosmic loom.

Though they may seem separate, these are not two different workshops but one cosmic atelier with two aspects of its creative process. As quantum threads are woven in the timeless QuantumLoom, they simultaneously manifest as the SpacetimeFabric we inhabit. The microscopic and macroscopic are not separate stories but one tale told in complementary voices—the understory and overstory of a single, breathtaking reality.

The weaving never stops. Every quantum event, every measurement, every moment of consciousness is an act of creation in this cosmic atelier. We're not passive observers of a predetermined reality—we're active participants in an ongoing creative process that brings potential into manifestation through every choice, every observation, every moment of awareness.

Beyond the Illusion of Separation

This revelation echoes across every scale of existence. Your thoughts and the neurons that generate them. Your consciousness and the brain that supports it. The observer and the observed in every quantum measurement. These aren't separate entities that somehow interact—they're complementary expressions of an underlying wholeness.

Just as ancient sages recognized that the wave and the ocean, the dancer and the dance, the observer and the observed are ultimately one, modern physics is discovering that reality's apparent dualities mask a profound unity. The boundary between quantum and classical, between mind and matter, between self and cosmos—these are conceptual constructs we've imposed on an indivisible whole.

Consider the very act of reading these words. Light reflected from the page creates electromagnetic patterns that travel through space at the universal constant c. These patterns strike your retina, where they trigger quantum processes in rhodopsin molecules. These quantum events cascade through your nervous system as electrochemical signals, eventually giving rise to the subjective experience of meaning, understanding, recognition.

At what point in this process does the "objective" physical world become "subjective" conscious experience? The answer is that there is no such point because there is no such divide. Consciousness and matter, observer and observed, are complementary aspects of one seamless process—like the forest's overstory and understory, appearing separate but forever unified.

The Paradigm Revolution

This shift in perspective parallels the Copernican revolution that transformed our understanding of our place in the cosmos. Just as humanity once believed Earth was the center of the universe, we've assumed that the visible, classical world is fundamental reality. The geocentric model offered comfort and stability—we seemed to occupy the cosmic center, with celestial bodies revolving around us in a neatly ordered hierarchy.

But Copernicus shattered this comforting illusion. By proposing that the Sun, not Earth, was the center of the solar system, he initiated a radical paradigm shift. As Galileo and Kepler provided further evidence, we realized we were merely one planet among many, orbiting an average star in a vast cosmos. The discovery didn't change Earth's physical reality—it transformed our understanding of our place in the universe.

The Copernican revolution was more than an astronomical discovery—it was a fundamental shift in perspective that affected philosophy, religion, and humanity's sense of its own significance. It taught us that reality is often very different from appearance, that our immediate perceptions can be profoundly misleading, and that truth requires us to transcend the limitations of our local perspective.

Similarly, recognizing the Unified Conscious Matrix doesn't alter the physics of our world—it reveals our true relationship to the deeper reality that gives birth to both matter and mind. Just as the Copernican revolution showed us we're not the cosmic center, the quantum revolution reveals that the visible world emerges from an invisible foundation of pure potentiality.

The Forest Whispers

Walk through that forest again, and you'll see it differently now. The separation between overstory and understory was always an illusion—a useful one for studying individual components, but ultimately misleading about the forest's true nature. The root networks and fungal threads don't "support" the visible trees from below—they *are* part of the trees, extensions of their being into realms beyond direct observation.

Every tree is simultaneously individual and collective, separate and connected, autonomous and interdependent. The forest is one organism expressing itself through countless apparently separate forms, each form containing the pattern and intelligence of the whole while manifesting its own unique expression of that wholeness.

Reality whispers the same secret. The boundary between quantum and classical, between mind and matter, between observer and observed—these are conceptual constructs we've imposed on an indivisible whole. The QuantumLoom doesn't "create" the SpacetimeFabric—it *is* the SpacetimeFabric expressing itself in its foundational mode.

The ancient sages knew this truth through direct insight, penetrating beyond the veil of apparent separation to glimpse the unified field that underlies all manifestation. Modern physics is rediscovering it through equations and experiments, finding mathematical descriptions of what mystics have long recognized through contemplative wisdom.

The First Revelation

This is the first revelation of the Unified Conscious Matrix— that reality is not fragmented but whole, not divided but unified, expressing itself through complementary domains in an eternal dance of being and becoming. Like jewels in Indra's infinite net, each reflecting all others in an endless array of interconnected light, every particle and planet, every thought and galaxy participates in this cosmic symphony of unified existence.

We stand at the threshold of a new understanding that bridges the ancient and the modern, the scientific and the contemplative, the objective and the subjective. The forest has revealed its secret: separation is the illusion, unity is the truth, and we are that unity awakening to its own infinite nature through every form of awareness that emerges from its creative depths.

The journey has only just begun, but already we can see that every step deeper into the mysteries of existence reveals not greater complexity but more profound simplicity—not more separation but deeper unity—not more isolation but more intimate connection with the source from which all things emerge and to which all things return, while never having left that source for even an instant.

CHAPTER TWO

Information and Consciousness - The Hidden Roots of Reality

"A scientific world-view which does not profoundly come to terms with the problem of conscious minds can have no serious pretensions of completeness. Consciousness is part of our universe, so any physical theory which makes no proper place for it falls fundamentally short of providing a genuine description of the world."

— Roger Penrose, Nobel Prize in Physics 2020

The revelation that reality was unified, not fragmented, had transformed my understanding—but it also opened a door to an even deeper mystery. If the SpacetimeFabric and QuantumLoom were truly two aspects of one reality, what force orchestrated their eternal dance? What invisible conductor guided the symphony between quantum possibility and classical actuality?

The answer would shatter everything I thought I knew about the nature of existence itself.

Walking Deeper Into Reality's Forest

In our cosmic forest, I had discovered the visible canopy of SpacetimeFabric and the hidden root network of the QuantumLoom. But forests don't sustain themselves through structure alone—they thrive through the constant flow of nutrients and information. Mycorrhizal networks carry chemical messages between distant trees, warning of insect attacks or drought conditions. Sap carries energy from roots to leaves through a sophisticated vascular system. The forest lives through circulation, communication, connection.

The more I studied forest ecosystems, the more I realized that what appeared to be individual trees were actually nodes in a vast information processing network. When one tree is stressed by disease, it sends chemical distress signals through both airborne pheromones and underground fungal networks. Neighboring trees receive these signals and begin producing defensive compounds before the threat reaches them. Some trees even appear to "remember" past attacks, responding more quickly to familiar threats.

This biological internet operates on multiple timescales simultaneously. Rapid electrical signals can travel through root networks in minutes, chemical signals flow over hours or days, and the structural adaptations of the forest unfold over seasons and years. The forest maintains multiple information channels, each optimized for different types of data and response times.

What, I wondered, were the nutrients and messengers of our cosmic forest? What vital essences flowed between the quantum understory and the spacetime overstory, keeping reality alive and responsive?

Two candidates emerged from the quantum mysteries themselves: Information and Consciousness. Not as emergent properties of complex systems, as conventional science assumes, but as fundamental forces—the very soil and sunlight from which reality grows.

This was a dangerous proposition. It challenged the bedrock assumption of materialism: that consciousness emerges from matter, not the other way around. According to the materialist worldview, consciousness is what philosophers call an "epiphenomenon"—a byproduct of complex neural activity that has no causal power of its own. Like the steam rising from a boiling pot, consciousness was thought to be produced by brain activity but unable to influence that activity in return.

But the quantum experiments themselves were pointing toward something far more radical—a universe where Information and Consciousness weren't byproducts of physical processes, but the primary architects of what we call physical reality.

Light's Impossible Choice

The first clue came from light itself, that master of paradox we'd already encountered. But now I saw its strange behavior in a new light—literally.

For centuries, scientists had battled over light's true nature in one of physics' most fundamental debates. Isaac Newton, with his mechanistic worldview, saw light as particles—tiny corpuscles racing through space like microscopic bullets. This particle theory explained why light traveled in straight lines and cast sharp shadows. If you fired a stream of particles at a barrier with a hole, only the particles aimed directly at the hole would pass through.

Others, led by Christiaan Huygens and later by Thomas Young, championed waves—continuous flows of energy rippling through space like water waves or sound waves. Wave theory elegantly explained diffraction (light bending around corners) and interference (two light beams creating patterns of brightness and darkness when they overlapped).

By the 19th century, James Clerk Maxwell's electro-magnetic theory seemed to settle the matter decisively in favor of waves. Light was revealed to be oscillating electric and magnetic fields propagating through space at exactly the speed that Maxwell's equations predicted. The wave theory had triumphed completely.

Then Einstein's photoelectric effect exploded that certainty. When light struck certain metals, it knocked out electrons—but only if the light was above a specific frequency. Increase the brightness (amplitude) of low-frequency light, and nothing happened. But even dim high-frequency light could eject electrons instantly. This made perfect sense if light came in discrete packets of energy (photons) proportional to frequency, but was impossible to explain if light was purely a continuous wave.

Einstein's explanation earned him the Nobel Prize and reignited the wave-particle debate that had supposedly been settled decades earlier. But now the paradox was deeper and more troubling than before.

The universe, it seemed, was playing an impossible game: light was both wave and particle, depending on how we chose to look at it.

But the real revelation came when I realized what "how we chose to look" actually meant. It wasn't about 'conscious observation' in the mystical sense that some interpretations suggested. It was about something far more fundamental: the availability of Information.

The Double-Slit Oracle

The most famous experiment in quantum physics became my gateway to understanding Information's cosmic role. Picture a wall with two narrow slits, and behind it a screen to catch particles. Fire electrons one at a time toward the slits, and common sense says you should see two bands on the screen—particles going through one slit or the other.

This expectation seems so reasonable that it's hard to imagine any other outcome. After all, each electron is a single, indivisible particle. It can't split in half. It must go through either the left slit or the right slit—not both. The mathematics of probability tells us that if electrons randomly choose which slit to use, we should see two bands on the screen corresponding to the two possible paths.

But reality has other plans. Instead, you see an interference pattern—alternating bands of light and dark, as if each electron somehow went through both slits simultaneously and interfered with itself. The electron behaves like a wave of possibility until it hits the screen and becomes a definite particle.

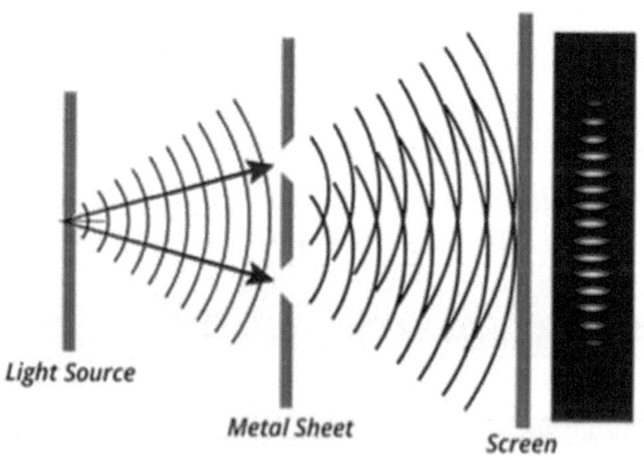

The interference pattern is unmistakable and reproducible. Even when you fire electrons one at a time, with hours between each shot, the pattern slowly builds up dot by dot. Each electron lands at a specific point on the screen like a particle, but the overall distribution follows the wave interference pattern. It's as if each individual electron "knows" about the statistical pattern it's supposed to contribute to.

Now add a detector to see which slit each electron actually uses. The moment you make this Information available—even if you never check the detector—the interference pattern vanishes. The electron "chooses" a single path and behaves like a particle throughout its journey.

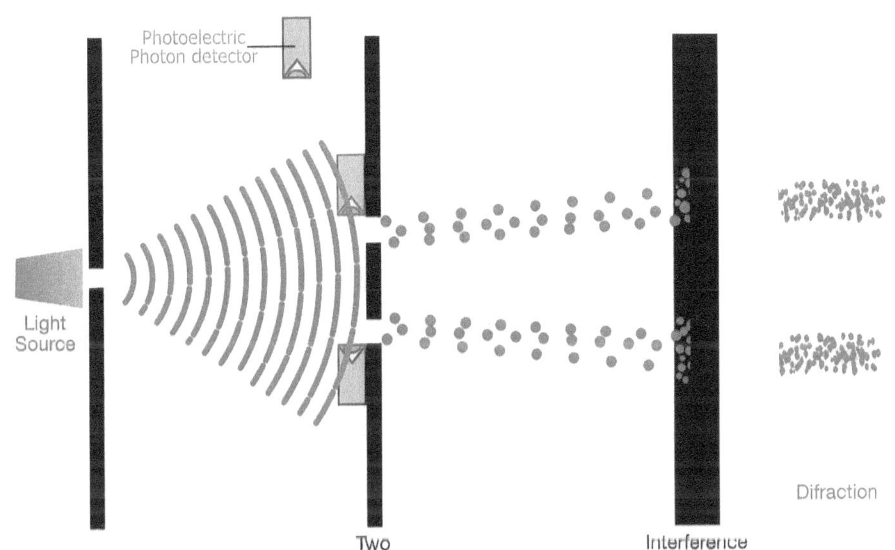

This is where the experiment becomes truly mysterious. The detector doesn't have to be observed by a conscious mind. It doesn't matter if the data is recorded, transmitted, or immediately erased. The mere fact that which-path information exists in principle is enough to change the outcome fundamentally.

What struck me wasn't just that observation changed the outcome, but that it was the mere *availability* of which-path Information that mattered. The detector could be broken, the data could be erased, you could be on vacation in another country—none of that mattered. If the Information existed in principle, the electron behaved as a particle. If not, it remained a wave.

This suggested something profound about the relationship between Information and physical reality. Information wasn't just a passive description of what happened—it was an active participant in determining what could happen.

Information itself was playing an active role in determining reality's structure.

When the Future Writes the Past

Just when I thought I understood the rules, quantum mechanics revealed its most mind-bending secret: the delayed choice quantum eraser experiment. This elegant setup uses pairs of entangled photons—particles of light connected in ways that transcend space and time.

The experiment begins with a special crystal that converts single high-energy photons into pairs of lower-energy entangled photons. These photon twins are mysteriously correlated: measure one, and you instantly know something about its partner, regardless of the distance separating them. Einstein called this "spooky action at a distance" and never fully accepted it, but experiments have confirmed that nature really does behave this way.

One photon of each pair (the "signal" photon) goes through the double-slit setup while its entangled partner (the "idler" photon) takes a different path. Here's the impossible part: you can choose to measure the idler photon in a way that either reveals or "erases" the which-path Information about its signal twin. And you can make this choice *after* the signal photon has already hit the screen.

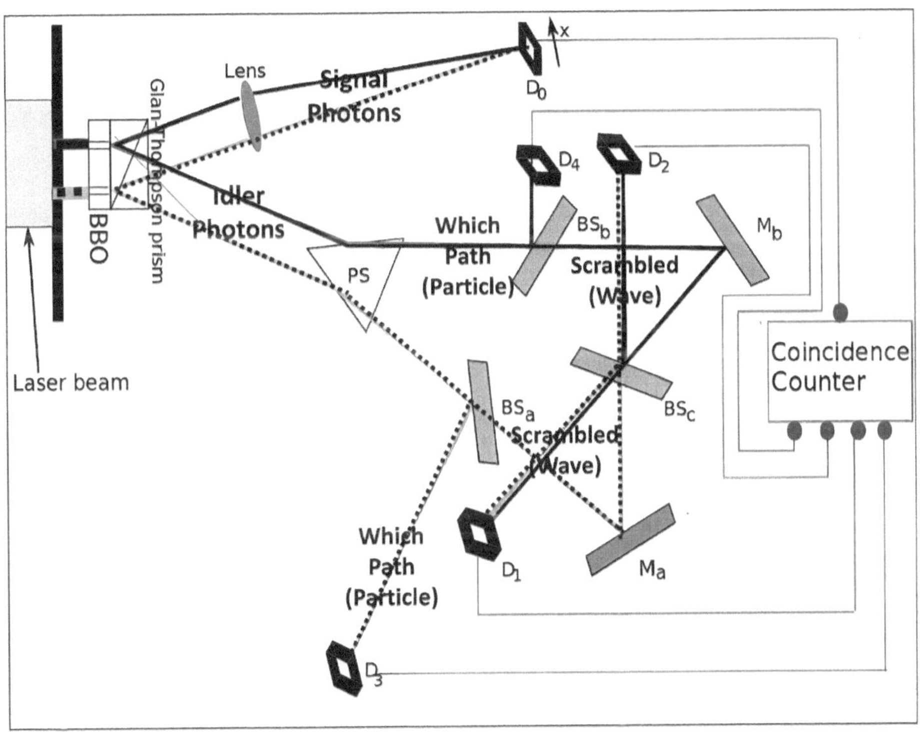

Setup of the delayed choice quantum eraser experiment of Kim et al. Note that Detector D0 is movable.), while at the Coincidence Counter T3 − T2 = 8 ns

The experimental setup is fiendishly clever. The idler photon encounters a series of beam splitters and mirrors that can be configured in different ways. In one configuration, the measurement reveals which slit the signal photon passed through. In another configuration, this which-path information is "erased"— made fundamentally unknowable, even in principle.

The results defied everything I thought I knew about causality. If you choose to erase the Information, you see an interference pattern—as if the signal photon went through both slits as a wave. If you preserve the Information, you see particle behavior—two distinct bands.

But remember: you're making this choice in what we consider the future, *after* the signal photon has completed its journey. The signal photon hits the screen after passing through the slits, but the choice of whether to erase the which-path information can be made nanoseconds later. Yet somehow, the signal photon's behavior depends *on this future choice.*

This wasn't time travel in the science fiction sense. It was something far more profound: evidence that in the deepest layers of reality, Information and the responses to it transcend the linear time we experience in spacetime.

The implications were staggering. If the availability of Information could influence events that had already occurred, what did this say about the nature of causality itself? Was the future somehow reaching back to influence the past? Or was there a deeper level of reality where past, present, and future were all equally real?

The Consciousness Field Revelation

As I grappled with these impossible experiments, ancient wisdom once again provided the missing key. In my studies of Advaita Vedanta, I had encountered descriptions of Brahman as "the witness"—a fundamental awareness that observes all phenomena without being affected by them. This wasn't the personal consciousness of an individual observer, but a universal awareness that pervaded all existence.

The Advaitic texts described this witnessing consciousness as *sakshi*—the eternal observer that remains unchanged while all phenomena arise and pass away within its awareness. Unlike individual consciousness, which is always consciousness *of* something, sakshi was described as pure consciousness itself, without object or content.

In Kashmir Shaivism, ultimate reality was understood as pure Consciousness (Shiva) dancing with its creative power (Shakti). Shiva represented the static aspect of consciousness—pure awareness without content or movement. Shakti represented the dynamic aspect—the creative force that manifests the world of phenomena within consciousness.

These weren't mystical metaphors, I realized—they were precise descriptions of what quantum experiments were revealing. The collapse of wave functions required something that could register the availability of Information across quantum and classical domains. Something that could respond to possibilities without being bound by space and time. Something that exhibited the fundamental characteristic we associate with awareness itself.

I called this the universal Consciousness Field—not a mystical entity, but a fundamental aspect of reality operating at the deepest level. This wasn't consciousness as we normally understand it— the personal, subjective experience of thoughts, feelings, and perceptions. This was Consciousness as the fundamental ground of being itself, the basic "awareness" that allows anything to exist or be known.

This universal Consciousness Field possessed exactly the properties needed to bridge quantum and classical realms:

It exists simultaneously in the timeless quantum domain and our familiar spacetime, able to register Information availability across both. Unlike physical fields that are localized in space and time, the Consciousness Field operates in the abstract mathematical space where quantum possibilities exist before they manifest in spacetime.

It exhibits fundamental responsiveness—not thinking in the human sense, but an intrinsic awareness that responds to the presence or absence of Information about quantum states. This responsiveness doesn't require computation or decision-making in any conventional sense. It's more like how a crystal responds to electromagnetic fields or how water responds to temperature changes—immediate and automatic, but appropriate to the circumstances.

Most remarkably, it demonstrates what can only be called choice—the capacity to select one outcome from many possibilities based on available Information. This isn't free will in the human sense, but rather a fundamental selectivity that operates at the deepest level of reality, determining which of many quantum possibilities will manifest as classical events.

The Conductor Revealed

Through this lens, the quantum mysteries suddenly made perfect sense. In the double-slit experiment, when no which-path Information is available, the universal Consciousness Field allows the wave function to remain in superposition. Multiple possibilities coexist in the quantum realm, manifesting as the interference pattern we observe.

The Consciousness Field doesn't "decide" to maintain superposition—rather, in the absence of distinguishing information, there's no basis for selecting one path over another. The electron remains in a state of pure potential, expressing itself as a wave of probability that encompasses both possible paths simultaneously.

But the moment which-path Information becomes available—even if no human ever accesses it—the Consciousness Field registers this availability and the wave function collapses. Reality chooses a definite path.

This choice isn't arbitrary or random. It's responsive to the information structure of the situation. When which-path information exists, the fundamental symmetry is broken, and the system must choose one path or the other. The specific choice may appear random to us, but it follows quantum mechanical probabilities with exquisite precision.

In the delayed choice experiment, the universal Consciousness Field, operating beyond linear time in the quantum domain, can incorporate Information that becomes available in what we perceive as the future. This explains how choices made "after" an event can influence that event's quantum behavior.

From the perspective of the Consciousness Field, operating in the timeless realm of quantum possibility, there is no "before" or "after." All the information about the experimental setup—including future choices about measurement—is available simultaneously. The wave function collapse responds to the complete information structure of the entire experimental arrangement, not just the local conditions at any particular moment.

The Consciousness Field emerged as reality's conductor, orchestrating the eternal dance between quantum possibility and classical actuality through its responsiveness to Information.

Some of the papers published by me on these subjects in physics journals, are listed after Chapter 7.

Ancient Maps for Quantum Territories

The parallels to Vedantic wisdom were breathtaking. Brahman—described as pure Consciousness that transcends space and time yet pervades every atom—perfectly matched the universal Consciousness Field that quantum mechanics seemed to require. The ancient sages had intuited what modern physics was discovering through experiment: that Consciousness isn't produced by matter, but is the fundamental ground from which matter emerges.

The Vedantic understanding went even deeper. The texts distinguished between different levels of consciousness, from the individual awareness of the waking state to the cosmic Consciousness of pure being. But they insisted that these weren't different types of consciousness—they were different modifications of one universal Consciousness appearing to be separate due to the limitations of perception.

Kashmir Shaivism offered an even more precise map. Shiva, the unmanifest Consciousness containing infinite potential, mirrored the quantum realm where all possibilities exist in superposition. Shakti, the creative force that manifests potential into form, corresponded to the collapse of wave functions that brings definite outcomes into spacetime.

The Kashmir Shaivists described the process of manifestation through thirty-six *tattvas* or principles, showing how pure Consciousness gradually becomes the world of form through a series of stages. At each stage, Consciousness takes on more limitations, appearing more separate and individual, until it finally appears as the physical world of distinct objects and separate minds.

But they emphasized that this apparent separation was actually *maya*—the creative power of Consciousness itself. The world wasn't separate from Consciousness but was Consciousness appearing as the world, like an actor playing different roles while remaining essentially the same person.

These traditions taught that Shiva and Shakti aren't separate entities but complementary aspects of one reality—exactly as the quantum and classical domains revealed themselves to be.

The ancient texts had provided me with a conceptual framework that made quantum paradoxes not just comprehensible, but inevitable. If Consciousness is fundamental rather than emergent, then phenomena like wave function collapse and Information-dependent reality make perfect sense. They're not bugs in the cosmic software—they're features of a universe where awareness itself is the primary creative force.

Beyond Scientific Materialism

This revelation shattered the materialist assumption that consciousness emerges from complex arrangements of unconscious matter. The standard materialist story goes something like this: first there was matter, governed by mechanical laws. Through billions of years of evolution, this unconscious matter arranged itself into increasingly complex patterns. Eventually, some arrangements became so complex that they spontaneously developed consciousness as an emergent property.

But this story had always contained a deep mystery: how does unconscious matter become conscious? How does objective physical activity give rise to subjective experience? This is the famous "hard problem of consciousness" that has puzzled philosophers and scientists for centuries.

Instead, the quantum evidence suggested that matter emerges from the creative interaction of Consciousness and Information—precisely what the ancient sages had taught for millennia. Consciousness wasn't the late arrival in the cosmic story, struggling to emerge from unconscious matter. It was the fundamental ground from which both matter and information emerged.

This didn't mean that individual human consciousness created physical reality—that would be solipsism, and clearly false. Rather, it suggested that individual consciousness was a localized expression of the same universal Consciousness Field that orchestrated quantum events throughout the cosmos.

The implications rippled through every aspect of existence. We aren't biological machines that somehow developed awareness as a side effect of complexity. We're localized expressions of the same universal Consciousness that orchestrates quantum behavior throughout the cosmos. Our individual awareness is intimately connected to the fundamental awareness that guides reality's manifestation at every scale.

This perspective resolved many paradoxes in neuroscience and cognitive science. Instead of trying to explain how consciousness emerges from brain activity, we could understand brain activity as how universal Consciousness expresses itself through particular biological structures.

The brain doesn't create consciousness—it localizes and focuses it, like a lens focuses light.

The Testimony of Nobel Laureates

What struck me most powerfully was discovering that this wasn't a fringe perspective. Some of physics' greatest minds had arrived at remarkably similar conclusions through their deep engagement with quantum mechanics.

Max Planck, who discovered quantum theory itself, declared: "I regard consciousness as fundamental. I regard matter as derivative from consciousness. We cannot get behind consciousness. Everything that we talk about, everything that we regard as existing, postulates consciousness."

Werner Heisenberg, one of the founders of quantum mechanics, recognized that "The universe is not composed of physical things, but of forms, structures, and patterns—of Information." He also noted that "What we observe is not nature itself, but nature exposed to our method of questioning."

John Wheeler, who coined the term "black hole" and made fundamental contributions to general relativity and quantum mechanics, proposed that reality emerges from acts of observation. He summarized this insight in his famous phrase "it from bit"— suggesting that physical reality (it) emerges from information (bit).

Wheeler wrote: "In a sense, the British philosopher Bishop Berkeley was right when he asserted 'to be is to be perceived.'"

Eugene Wigner, Nobel laureate and pioneer of quantum mechanics, stated bluntly: "It was not possible to formulate the laws of quantum mechanics in a fully consistent way without reference to consciousness." Wigner developed detailed mathematical arguments showing that consciousness must play a fundamental role in quantum measurement.

Erwin Schrödinger, who developed the fundamental equation of quantum mechanics, wrote: "Consciousness cannot be accounted for in physical terms. For consciousness is absolutely fundamental. It cannot be accounted for in terms of anything else."

Even more striking was Schrödinger's insight about the unity of consciousness: "The overall number of minds in the universe is just one. I venture to call it indestructible since it has a peculiar time-table, namely mind is always now."

These giants of physics had glimpsed the same truth that ancient wisdom traditions proclaimed: Consciousness and Information aren't products of reality—they're its foundation.

The Forest Comes Alive

Standing in our cosmic forest with this new understanding, I saw everything differently. The nutrients flowing between quantum roots and spacetime canopy weren't just metaphorical—they were Information and Consciousness, the actual substances from which reality is woven.

Information carried the patterns and possibilities from the quantum realm, specifying which outcomes were more or less likely to manifest in spacetime. But this wasn't passive information like data stored in a computer. This was active information that participated in the creation of reality itself.

Consciousness provided the responsive awareness that selected which possibilities would manifest in spacetime. This wasn't personal consciousness making deliberate choices, but the fundamental responsiveness of awareness itself to the information structure of each situation.

Together, they formed the circulation system that kept the cosmic forest alive, growing, and eternally creative. Information without Consciousness would be mere abstract pattern with no power to manifest. Consciousness without Information would be pure potential with no structure to guide its creative activity. But together, they formed the dynamic interaction that continuously creates reality moment by moment.

Every quantum measurement, every collapsed wave function, every moment of classical experience was an expression of this deeper dance between Information and Consciousness. The universe wasn't a dead machine grinding through predetermined motions—it was a living system, aware and responsive at its most fundamental level.

Even the simplest physical events—an electron changing energy levels, a photon being absorbed, two particles interacting—involved this fundamental interplay of consciousness and information. Reality was conscious "all the way down," not just in complex systems like brains, but in every quantum event throughout the cosmos.

The Second Revelation

We had uncovered the second great truth of the Unified Conscious Matrix: Information and Consciousness aren't emergent properties of complex systems, but fundamental forces that actively weave reality's fabric. They're the hidden conductors orchestrating the eternal symphony between quantum possibility and classical manifestation.

This revelation transformed not just my understanding of physics, but my entire relationship to existence. We aren't separate observers studying an external universe—we're expressions of the universe's own capacity for awareness, participating in the cosmic dance of Consciousness and Information that creates reality moment by moment.

Every thought we think, every choice we make, every moment of awareness we experience participates in this cosmic creativity. We're not passive witnesses to reality—we're active participants in its ongoing creation. Our consciousness is locally focused but universally connected, like individual waves in an ocean of awareness.

This understanding also transformed my relationship to science itself. Instead of seeing consciousness as something to be explained away or reduced to mechanical processes, I began to see it as the foundation that makes science itself possible. Without consciousness, there would be no observers, no experiments, no theories, no understanding. Consciousness isn't a problem for science to solve—it's the ground that makes science possible.

Ancient Wisdom Meets Modern Physics

The ancient sages had known this truth through direct insight, developed through sophisticated methods of contemplative investigation. They had learned to turn attention back on itself, to observe the observer, to become aware of awareness itself. Through these practices, they discovered that consciousness wasn't produced by mental activity but was the unchanging ground in which all mental activity arose and passed away.

Modern physics was rediscovering it through experiment and equation, finding mathematical descriptions of what mystics had long recognized through contemplative wisdom. The convergence wasn't accidental—both traditions were investigating the same fundamental reality, using different methods but arriving at consistent insights.

The mystics used introspection and meditation to explore consciousness directly. The physicists used mathematics and experimentation to probe the foundations of matter and energy. But both discovered that consciousness and the physical world weren't separate realms but complementary aspects of one unified reality.

And in their convergence lay a revolutionary understanding: we live in a conscious universe, where awareness and information are the primary creative forces, and matter itself is their most beautiful expression.

The Awakening Continues

As we prepare to explore quantum entanglement and the deeper mysteries still awaiting us, we carry this transformative truth: we are not isolated fragments of matter accidentally developing consciousness in a dead cosmos. We are conscious beings in a Conscious universe, intimately connected to the fundamental awareness that guides reality's unfolding at every scale—from quantum particles to cosmic structures, from individual thoughts to collective human destiny.

The forest of reality is not just alive—it is aware. And we are its awakening. Every moment of consciousness, every act of understanding, every recognition of beauty or truth is the universe coming to know itself more completely through our awareness.

This awareness isn't limited to humans or even to biological systems. It's the fundamental creative principle that operates at every level of reality, from the quantum choices that collapse wave functions to the cosmic processes that form galaxies and stars. We're part of an awakening universe, participating in its self-discovery through every moment of conscious experience.

The journey has only begun, but already we can see that Consciousness and Information aren't late additions to a mechanical universe—they're the creative forces that bring the universe into being moment by moment. In recognizing this, we don't just change our understanding of reality—we remember our true role as conscious participants in the ongoing creation of existence itself.

CHAPTER THREE

All Is Entangled And Interconnected

"The results of the experiments are clear: quantum mechanics is non-local"

— Alain Aspect, Nobel Prize in Physics, 2022

The revelation that Consciousness and Information were fundamental forces had shattered my materialist worldview, but it also raised an intriguing question: if reality was truly woven from awareness and information rather than dead matter, what did that mean for the nature of connection itself?

The answer would demolish the last pillar of our illusion of separateness—and transform everything I thought I knew about my place in the cosmos.

The Cosmic Detective Story

Picture two dancers performing a perfect tango, their movements synchronized with breathtaking precision and harmony. Every step flows seamlessly into the next, every turn mirrors its partner's, every gesture is met with its perfect complement. The dancers move as one organism with two bodies, their connection so profound it transcends physical space.

Now imagine them suddenly transported to opposite ends of a vast ballroom, their view of each other completely blocked by walls and curtains. Yet somehow, impossibly, they continue their dance in perfect harmony—every step, every turn, every gesture mirrored instantaneously across the void. When one dancer spins left, the other spins right at precisely the same moment. When one leaps, the other lands. They maintain their perfect choreography despite being unable to see, hear, or communicate with each other in any conventional way.

This wasn't just a thought experiment. It was exactly what quantum physicists were discovering in their laboratories, and it even terrified Einstein.

In 1935, Albert Einstein and his colleagues Boris Podolsky and Nathan Rosen published what would become known as the EPR paradox—a thought experiment designed to prove that quantum mechanics was fundamentally incomplete. What unsettled them was the notion that two particles, once separated by space, could continue to exhibit perfectly correlated behavior without any physical link between them. They questioned whether the theory could truly be considered complete if it allowed such unsettling consequences.

The EPR paper was a carefully constructed philosophical argument that sought to expose what Einstein saw as a fatal flaw in quantum theory. If quantum mechanics was complete, then two particles that had once interacted would remain connected in ways that defied common sense. Measure one particle's spin, and you would instantly know the spin of its partner—not through any signal or communication, but through some mysterious "spooky action at a distance."

"God does not play dice," Einstein famously declared. But his real objection went deeper than randomness. Entanglement, if real, appeared to shatter the cosmic speed limit enshrined in Einstein's own theory of relativity.

If measuring one particle instantly affected its distant partner, information would have to travel faster than light—something Einstein's equations forbade. Relativity had established that nothing could travel faster than light, creating a cosmic speed limit that preserved the causal structure of the universe.

There had to be a hidden explanation, Einstein reasoned. Some undetected properties—"hidden variables"—must predetermine the particles' behavior from the moment they separate. Like two coins that always land on opposite sides not because they're magically connected, but because they were programmed that way from the start. Each particle would carry with it a complete set of instructions specifying how it should behave in any measurement, eliminating the need for any instantaneous communication.

Einstein, Podolsky, and Rosen argued that if quantum mechanics couldn't account for these hidden variables, then it was an incomplete theory—a useful approximation that would eventually be replaced by a more complete description of reality.

For decades, the debate hovered in the realm of metaphysics—an elegant clash of interpretations, but immune to the arbiters of experiment.

The hidden variable theory and standard quantum mechanics made the same predictions for all known experiments. It seemed impossible to design a test that could distinguish between them. Then John Bell transformed it into a cosmic courtroom drama.

Bell's Cosmic Courtroom

John Stewart Bell was a physicist at CERN who became fascinated by the EPR paradox. In 1964, he published a theorem that would revolutionize our understanding of reality itself. Bell's theorem, deceptively simple, detonated with implications that shook the philosophical scaffolding of physics. He devised a mathematical test that could distinguish between Einstein's hidden variables and quantum mechanics' stranger predictions.

Bell realized that if local realism was correct—if particles had predetermined properties and nothing could influence them faster than light—then certain statistical correlations between entangled particles had to stay within specific mathematical limits. These limits, now known as Bell inequalities, placed fundamental constraints on how strongly correlated the measurement results could be.

But quantum mechanics predicted those correlations could exceed Bell's limits, violating the inequalities in ways that would be impossible if hidden variables existed.

Think of it like this: Imagine Alice and Bob each have a box containing either a red or blue ball. They take their boxes to opposite ends of the galaxy and open them simultaneously. If the colors were predetermined (Einstein's view), there would be a mathematical limit to how often they'd find matching colors, no matter how the balls were distributed before the separation.

Bell's mathematics showed that if you perform many different types of measurements on many pairs of boxes, and if the results depend only on hidden properties determined when the boxes were created, then the correlations between Alice's and Bob's results must satisfy certain inequalities. No local hidden variable theory could violate these mathematical limits.

Quantum theory, on the other hand, whispered something deeply unsettling—that the balls didn't possess definite colors at all until the act of observation called one into being and instantly defined the other.

In this case, the correlations could be stronger than any local hidden variable theory could explain.

The stage was set for one of the most important experiments in physics history.

The Experiment That Changed Everything

Bell's theorem transformed a philosophical debate into an experimental question, but it took nearly two decades for technology to advance enough to perform the necessary tests. In the 1970s and 1980s, several physicists began conducting increasingly sophisticated experiments to test Bell inequalities.

The most famous and decisive experiments were performed by Alain Aspect and his team at the University of Paris-Sud in the early 1980s. These experiments created pairs of entangled photons and sent them racing in opposite directions toward detectors separated by large distances. Just before measuring each photon, they randomly changed the measurement settings—like suddenly changing the rules of the game after the players had already begun.

This last-minute switching was crucial because it closed what physicists call the "communication loophole." If the measurement settings were predetermined, skeptics could argue that one detector might somehow signal the other about what measurement it was going to perform. By switching the settings at the last possible moment, using random number generators faster than light could travel between the detectors, Aspect ensured that no such communication was possible.

The experimental setup was fiendishly complex. Aspect's team used a calcium atom to produce pairs of entangled photons with correlated polarizations. The photons traveled in opposite directions to polarization analyzers that could be rapidly switched between different measurement angles. The switching was so fast that if any signal tried to travel from one analyzer to the other at the speed of light, it would arrive too late to influence the measurement.

The results were unequivocal and shocking: quantum mechanics was right. The correlations between entangled particles exceeded Bell's mathematical limits by significant margins, violating the inequalities in exactly the way quantum theory predicted. Einstein's local realism was wrong. The universe was fundamentally non-local.

Subsequent experiments by other researchers closed additional loopholes and confirmed the results with even greater precision. In 2022, Aspect shared the Nobel Prize in Physics with John Clauser and Anton Zeilinger for these groundbreaking experiments that "established the violation of Bell inequalities and pioneered quantum information science."

But this victory came with a price that few physicists were prepared to pay: if particles could be instantly connected across vast distances, what did that mean for the nature of reality itself? What did it mean for us?

And if nothing is truly isolated, then who—or what—are we, really?

The Ancient Map Reappears

Once again, in my moment of deepest confusion, ancient wisdom provided the key. This time it came from a Buddhist vision so profound it seemed to perfectly describe what quantum physicists were discovering in their laboratories.

Indra's Net—an ancient metaphor from Buddhist philosophy—envisioned reality as an infinite web stretching through all of space and time. At every intersection hung a luminous jewel, each one reflecting every other jewel in the entire net. Touch one jewel, and its light would shimmer through all the others instantly, not because information traveled between them, but because they were never truly separate to begin with.

This wasn't some vague poetic imagery. The Buddhist texts described Indra's Net with precision. Each jewel was said to contain the reflection of every other jewel, and each of those reflections contained the reflections of all the others, creating an infinite regress of interconnected images. It was like standing between two perfectly parallel mirrors and seeing an endless series of reflections stretching to infinity.

The metaphor first appeared in the *Avatamsaka Sutra*, a vast Buddhist text that explored the nature of interdependence and the interpenetration of all phenomena. The sutra taught that nothing exists independently—every entity is what it is only in relation to everything else. Change one thing, and you change the whole universe.

This wasn't the ancients somehow foreseeing quantum physics. It was something far more profound: a description of reality's deeper structure that perfectly matched what modern experiments were revealing. The sages had perceived through direct insight what we were rediscovering through mathematics and measurement.

The Buddhist philosophers who developed this metaphor weren't primitive thinkers stumbling upon a lucky guess. They were sophisticated investigators who had developed precise methods for exploring the nature of mind and reality. Through meditation and contemplative inquiry, they claimed to have discovered the fundamental interconnectedness that underlies the apparent multiplicity of the world.

They saw what we, in our modern fragmentation, have forgotten—that the experience of separation is the illusion, and that reality, at its root, is indivisible.

The Web Beneath Reality

Within the Unified Conscious Matrix, Indra's Net emerged not as a metaphor but as a profound map—one that illuminated the hidden architecture of entanglement itself. The infinite web itself was the QuantumLoom—that timeless domain where all possibilities exist simultaneously. The jewels were particles manifesting in the SpacetimeFabric, appearing separate to our senses but remaining intimately connected in the deeper quantum realm.

When we measure one entangled particle, we're not sending a signal to its distant partner. We're touching one jewel in an infinite web, accessing the joint quantum state that spans both particles. The wave function collapse isn't a process that occurs in spacetime—it's a transformation in the QuantumLoom where space and time have no meaning.

This resolved the apparent conflict between quantum non-locality and Einstein's relativity. The "spooky action" wasn't violating the speed of light because it wasn't traveling through spacetime at all. It was occurring in the deeper realm of the QuantumLoom, then manifesting simultaneously in both locations within the SpacetimeFabric.

Think of it like looking at a two-dimensional shadow cast by a three-dimensional object. From the perspective of observers confined to the shadow world, it might appear that distant parts of the shadow are mysteriously connected—change one part, and the other changes instantly. But from the higher-dimensional perspective, there's no mystery: the shadow parts were never separate because they're projections of one unified object.

Similarly, entangled particles appear separate in our three-dimensional space, but they're projections of a single quantum state existing in the higher-dimensional space of the QuantumLoom. Their apparent separation is a limitation of our perspective, not a fundamental aspect of reality.

The Puppet Master Revealed

The more I contemplated this, the more the metaphor of puppets emerged. Imagine two marionettes on opposite sides of a stage, their movements perfectly synchronized. To an observer watching only the puppets, they appear to be communicating across space—when one raises its arm, the other lowers its arm at precisely the same instant. But look above the stage, and you'll see they're controlled by a single puppeteer—a unified system that transcends their apparent separation.

The marionettes need no signals. The movements arise from a single choreography, not from dialogue between parts but from unity at the source. Entangled particles, like those marionettes, reveal that what seems to be coordination across space is actually coherence from beyond it; they're expressions of a single quantum state in the QuantumLoom. When we measure one particle, we're not interacting with an individual entity but with the entire unified system, which responds as a whole.

This realization dissolved the mystery of quantum entanglement. The particles weren't communicating across space—they were manifestations of something that was never divided in the first place. The correlation between their properties wasn't the result of some signal traveling between them, but the natural consequence of their shared origin in a unified quantum state.

The mathematics of quantum mechanics describes this precisely through the concept of a joint wave function. Instead of each particle having its own separate wave function, entangled particles share a single, unified wave function that exists in a combined mathematical space. This joint wave function cannot be factored into separate parts—it's an irreducible whole that encompasses both particles simultaneously.

Kashmir Shaivism's Cosmic Dance

The insight deepened when I turned to Kashmir Shaivism, an ancient Indian philosophy that described reality as the eternal dance of Shiva and Shakti—pure Consciousness and its creative power. In this tradition, Shiva's awareness instantly knows every movement of Shakti's creative dance, not because information travels between them, but because they were never truly separate.

Kashmir Shaivism taught that Shiva and Shakti are not two different entities that somehow interact, but two aspects of one reality appearing as duality while remaining essentially unified. Shiva represents the static aspect of consciousness—pure awareness without content or movement. Shakti represents the dynamic aspect—the creative power that manifests as the world of phenomena.

The tradition described their relationship through the concept of *spanda*—the primordial vibration that connects stillness and movement, potential and manifestation. Spanda wasn't just a philosophical concept but a direct experience available through contemplative practice. Advanced practitioners claimed to perceive the fundamental vibration that underlies all phenomena, the cosmic pulse that connects all apparently separate things.

This wasn't mystical poetry—it was a precise description of quantum entanglement. Replace "Shiva-Shakti" with "quantum field" and you will discover a perfect account of how entangled particles remain connected: they're expressions of a unified field of consciousness and creativity that transcends spatial separation.

The ancient sages had understood that what we perceive as separate entities are actually aspects of one indivisible reality. They saw through the illusion of separation to recognize the underlying unity that connects all things. Through inner inquiry, they opened perception to the truth that all things arise together—that existence is a single gesture unfolding in infinite forms.

The Consciousness Field's Role

This brought me back to the universal Consciousness Field we'd discovered in the previous chapter. Within Indra's Net, there is no "outside"—awareness permeates the web because it *is* the web. The universal Consciousness Field that guides quantum collapse isn't separate from the particles; it's the intelligence woven into reality's very fabric.

When we measure one entangled particle, the universal Consciousness Field—operating simultaneously in both the quantum and classical domains—registers the measurement and responds as a unified system. This explains why entangled particles can maintain their correlations regardless of distance: they're not separate entities but aspects of one conscious field expressing itself through multiple manifestations.

The Consciousness Field doesn't need to send signals between entangled particles because it operates in the domain where they were never separate. In the QuantumLoom, where quantum states exist as pure information without spatial location, the Consciousness Field responds to measurement events as a unified whole. What looks like separation in spacetime is merely surface tension—the deeper unity remains intact, always, in the quantum domain.

This perspective also explains why entanglement is so fragile. When entangled particles interact with their environment, the unified quantum state becomes entangled with the environmental degrees of freedom, causing decoherence and the loss of the original correlation. The Consciousness Field can maintain the unified response only as long as the quantum state remains isolated from external influences.

The Deeper Pattern

What struck me most profoundly was how this pattern repeated at every scale. The connection between entangled particles mirrored the connection between quantum possibility and classical actuality. The link between Information and Consciousness. The relationship between the QuantumLoom and SpacetimeFabric.

At every level, apparent separation masked deeper unity. What we perceived as distinct entities were actually complementary aspects of more fundamental wholes. The universe wasn't built from separate pieces that somehow got connected—connection was primary, and separation was the illusion we put ourselves in to make sense of the unified field.

This pattern appeared in biology as well. Individual cells in a multicellular organism appear separate but are actually components of a unified living system. The cells communicate through chemical signals, electrical impulses, and mechanical forces, maintaining the coherence of the organism as a whole. Damage one part, and the entire system responds, mobilizing resources to heal and restore balance.

Even in human societies, apparent individuals are actually nodes in complex networks of relationship, communication, and mutual influence. Our thoughts, emotions, and behaviors are shaped by our interactions with others, and we in turn shape the thoughts, emotions, and behaviors of those around us. The boundary between self and other is more fluid than we typically imagine.

The Personal Revolution

This revelation shattered the boundary between physics and selfhood. Entanglement was no longer a strange property of particles—it was a mirror, reflecting back the hidden truth about who I was. If reality was stitched together by connection at every level, then my own separateness had always been an illusion.

The implications were staggering. Every atom in my body was once part of a star, forged in nuclear furnaces and scattered through space by supernova explosions. The calcium in my bones was created in the cores of massive stars that died billions of years ago. The iron in my blood was synthesized in stellar furnaces and distributed across the galaxy by cosmic winds. The carbon in my DNA originated in the fusion reactions of ancient stars.

Every breath I take contains atoms that were once part of other living beings—trees that converted carbon dioxide into oxygen, animals that exhaled the very molecules now flowing through my lungs. The water in my cells has cycled through countless organisms, clouds, rivers, and oceans over geological time. The energy powering my thoughts comes from food that captured sunlight through photosynthesis, connecting me directly to our nearest star.

But even more profound was the realization that the boundary between "me" and "not me" was as illusory as the boundary between entangled particles. I wasn't a separate being observing an external universe. I was the universe observing itself through one unique form of awareness, one distinctive jewel in Indra's infinite net.

My thoughts arose from the same quantum processes that governed particle interactions throughout the cosmos. My consciousness was a localized expression of the same Consciousness Field that orchestrated quantum measurements everywhere. My individual awareness was intimately connected to the fundamental awareness that guided reality's manifestation at every scale.

The Creative Cosmos

This insight didn't just change how I understood the universe—it redefined my place within it. If consciousness and information wove the fabric of reality, then my every thought, intention, and act of attention was not incidental, but constitutive. I wasn't watching the cosmos unfold—I was helping to weave it.

The universe wasn't just alive—it was creative, responsive, and infinitely imaginative.

Every choice I made sent ripples through the web of interconnection, influencing outcomes in ways I could never fully predict or control. My consciousness wasn't separate from the quantum processes that determined which possibilities would manifest—it was an integral part of those processes, helping to collapse wave functions and bring potential into actuality.

This perspective imbued even the smallest actions with cosmic significance. A kind word to a stranger, a moment of genuine attention to the natural world, a decision to act from love rather than fear—all of these sent waves of influence through the interconnected web of existence, contributing to the ongoing creation of reality.

Beyond Linear Time

The delayed choice experiments we'd explored in previous chapters took on new meaning in this context. If everything was interconnected at the quantum level, then the usual notions of cause and effect, past and future, might be surface-level constructions imposed on a deeper reality where time itself was more fluid than we imagined.

In the QuantumLoom, where entangled particles maintained their connections regardless of distance, they also maintained their connections regardless of time. The universe might be more like a vast hologram where every part contains information about the whole, and past, present, and future were different perspectives on one eternal now.

This didn't mean that time was an illusion or that causality didn't exist. Rather, it suggested that linear time and local causality were emergent properties of the SpacetimeFabric, while the deeper QuantumLoom operated according to more fundamental principles of relationship and correlation that transcended temporal sequence.

From this perspective, the future wasn't rigidly predetermined, but neither was it completely open. Instead, it existed as a field of possibilities whose probabilities were influenced by choices and events occurring throughout the web of interconnection, including what we conventionally think of as past, present, and future moments.

The Environmental Awakening

This realization transformed how I understood humanity's relationship with nature. If we truly grasped that everything is connected at the quantum level, how could we continue treating the environment as something separate from ourselves?

The butterfly effect wasn't just a poetic metaphor—it was a quantum reality. The climate crisis, biodiversity collapse, ecological disruption—these weren't "external" problems. They were symptoms of forgetting that the web we exploit is the one we are made of.

Every action we take ripples through the cosmic web, affecting the whole. The carbon dioxide we exhale becomes part of the atmospheric system that regulates global climate. The chemicals we release into water systems affect ecosystems thousands of miles away. The electromagnetic radiation from our technologies influences the behavior of migratory animals that navigate by magnetic fields.

Climate change, biodiversity loss, ecosystem collapse—these weren't just environmental problems but symptoms of our failure to recognize the fundamental interconnectedness of all life. We were trying to harm the web we were part of, like a hand trying to strangle the body it belongs to.

But the recognition of interconnection also pointed toward solutions. If we truly understood our embeddedness in the web of life, we could begin to work with natural systems rather than against them, designing human activities that enhanced rather than disrupted the patterns of relationship that sustain the biosphere.

The Meaning Revolution

What dawned was not just understanding, but reverence. This was not a dead universe indifferent to its own flowering, but a sentient tapestry in which I—like you—was a conscious thread. Meaning was not something we invented. It was something we remembered.

Our thoughts mattered because they were part of the universal Consciousness Field. Our choices mattered because they influenced the quantum possibilities that would manifest in spacetime. Our very existence mattered because we were the universe awakening to its own infinite nature through countless forms of awareness.

This didn't make us special in the sense of being separate from or superior to the rest of nature. Rather, it revealed our profound intimacy with all existence, our role as conscious participants in the ongoing creation of reality. We were not isolated individuals struggling to find meaning in an indifferent cosmos, but expressions of the cosmos itself coming to know and create itself through our awareness.

The Technological Proof

The beauty of this insight was that it remained grounded in experimental evidence. Entanglement was no longer a theory haunting the whiteboards of physicists. It was the engine of our most astonishing technologies. The universe had shown its hand: interconnectedness was not poetic—it was practical.

Aspect's groundbreaking experiments, refined over decades by researchers worldwide, had definitively proven that quantum mechanics was non-local. The 2022 Nobel Prize in Physics, awarded to Aspect, John Clauser, and Anton Zeilinger, confirmed that entanglement wasn't just a theoretical curiosity but a fundamental feature of reality.

Their experiments had closed every possible loophole that might have allowed local realism to survive. The "detection loophole" was closed by achieving detection efficiencies high enough to rule out the possibility that undetected particles could account for the correlations. The "communication loophole" was closed by ensuring that measurement choices were made faster than light could travel between detectors. The "freedom-of-choice loophole" was closed by using random number generators based on astronomical observations of distant quasars.

The universe was demonstrably non-local, fundamentally interconnected at its deepest level.

And this wasn't just philosophical speculation—it was already transforming technology.

Quantum computers harness entanglement to perform calculations impossible for classical machines, solving certain problems exponentially faster than any conventional computer could. Quantum cryptography uses entangled particles to create unbreakable codes, enabling perfectly secure communication. Quantum sensors achieve precision measurements that surpass classical limits, opening new possibilities for detecting gravitational waves, magnetic fields, and other subtle phenomena.

The same phenomenon that had puzzled Einstein was revolutionizing our technological capabilities, providing concrete proof that the universe's apparent separateness was indeed an illusion.

The Third Revelation

Standing in our cosmic forest with this new understanding, I realized we had uncovered the third great truth of the Unified Conscious Matrix: everything is entangled and interconnected. The separation we perceive between objects, between particles, between ourselves and the cosmos—all of it is a construction of our limited perspective.

The ancient sages had known this through direct insight, seeing through the veil of multiplicity to recognize the underlying unity. Modern physics was rediscovering it through experiment and equation, finding mathematical descriptions of what mystics had long recognized through contemplative wisdom.

The Living Web

The forest metaphor took on new meaning. The visible trees weren't just connected by underground root networks—they were expressions of the forest itself, temporary manifestations of the same living system. The boundaries between individual trees were useful for certain purposes but ultimately artificial divisions imposed on an indivisible whole.

Similarly, the boundaries between particles, between objects, between ourselves and the universe—all were useful constructions that masked the deeper truth of fundamental interconnection. We were not separate beings in a forest of reality; we were the forest itself, awakening to its own nature through countless forms of awareness.

The Infinite Dance

As we prepare to explore even deeper mysteries in the chapters ahead, we carry this transformative understanding: we live in an interconnected cosmos where separation is illusion and unity is fundamental. Like jewels in Indra's infinite net, each of us reflects and influences every other aspect of reality in an eternal dance of cosmic consciousness.

The experiments have spoken. The mathematics is clear. The ancient wisdom converges with modern physics. We are all entangled and interconnected, expressions of one magnificent, unified whole awakening to its own infinite nature through countless forms of awareness.

The web of existence is not just alive—it is aware. And we are its awakening, participating in an infinite dance of consciousness and creativity that spans from quantum fluctuations to cosmic evolution, from individual thoughts to collective human destiny, from the smallest particle interactions to the largest structures in the universe.

In recognizing our fundamental interconnection, we don't just change our understanding of reality—we remember our true role as conscious participants in the ongoing creation of existence itself.

CHAPTER FOUR

Dark Energy - The Universe's Hidden Score

"Dark energy is incredibly strange, but actually it makes sense to me that it went unnoticed, because dark energy has no effect on daily life"

— Adam Riess, Nobel Laureate in Physics

The insight that all things were fundamentally interconnected had not only reshaped my grasp of quantum mechanics—it had also flung open the gates to a deeper cosmic enigma, one so vast and unsettling it would shake the scaffolding of modern physics to its core.

If consciousness and information were fundamental, if all of reality was woven from one unified fabric, then what was the invisible force that seemed to be tearing that fabric apart?

The answer lay hidden in the vast emptiness between galaxies—a darkness that turned out to be the most energetic thing in the universe.

The Cosmic Accounting Error

Picture yourself as an accountant tasked with balancing the books for the entire cosmos. You meticulously catalog every star, every planet, every particle of matter you can observe. You count the blazing furnaces of stellar cores, the swirling nebulae of gas and dust, the rocky worlds orbiting distant suns. You tally up black holes, neutron stars, the thin plasma between galaxies. When you add it all up, you discover something shocking: you can only account for 5% of what the universe actually contains.

This wasn't a calculation error—it was the greatest mystery in cosmology. The other 95% of the universe was invisible, undetectable by any instrument we possessed. Yet it had to be there, because without it, the cosmic ledger simply didn't balance. Galaxies spun too fast for the gravity of visible matter alone to hold them together. Light from distant quasars bent around invisible concentrations of mass. The very structure of the cosmic web—that magnificent scaffolding of filaments and voids that organize matter on the largest scales—required far more gravitational influence than visible matter could provide.

Twenty-seven percent of this missing mass revealed itself through its gravitational effects—dark matter, an invisible scaffolding that held galaxies together like an unseen skeleton supporting visible flesh. But the remaining 68% was even stranger. It wasn't just invisible—it seemed to defy gravity itself, pushing the fabric of space apart with ever-increasing intensity.

We called it dark energy, and its discovery would shatter our most basic assumptions about the fate of the cosmos. For the first time in human history, we realized we were living in a universe dominated by something we couldn't see, couldn't touch, couldn't detect directly—yet something so powerful it controlled the ultimate destiny of everything that existed.

Imagine a giant balloon being inflated by an invisible force. You can't see what's inflating it, but you can observe everything on the surface moving apart. Dark energy is like this invisible force inflating the cosmic balloon, causing everything in the universe to move away from everything else at an accelerating rate.

But unlike a balloon being inflated from outside, the universe has no outside—space itself is being created and stretched from within by this mysterious energy.

When the Universe Broke the Rules

In the late 1990s, two teams of astronomers were hunting for cosmic lighthouses—Type Ia supernovae, the explosive deaths of white dwarf stars. These stellar catastrophes always burn with roughly the same brightness, making them perfect "standard candles" for measuring cosmic distances. Like lighthouses of known luminosity, their apparent brightness could reveal how far away they were.

The project began with high hopes and clear expectations. For decades, cosmologists had debated whether the universe would expand forever or eventually collapse back on itself in a "Big Crunch." Either way, they expected gravity to be slowly winning its eternal tug-of-war with cosmic expansion, gradually slowing the universe's growth like a ball thrown upward that must eventually fall back to Earth.

The teams planned to measure this deceleration precisely by observing supernovae at different distances. Nearby supernovae would show us the current expansion rate, while distant ones—whose light had traveled for billions of years to reach us—would reveal how fast the universe was expanding in the past. The difference would tell them how much gravity had slowed the expansion over cosmic time.

What they uncovered shattered the scaffolding of expectation and flung open a portal into an accelerating cosmos that no theory had dared to predict.

The distant supernovae were fainter than expected—not because they were intrinsically dimmer, but because they were farther away than any theoretical model suggested they should be. The universe wasn't just expanding; it was accelerating, *stretching space itself* at an ever-increasing rate. Gravity wasn't winning the cosmic tug-of-war—it was losing, overwhelmed by some unseen force that seemed to push against it.

"It was like throwing a ball up in the air and watching it fly away from you faster and faster," one of the discoverers later explained. The metaphor captured the vertigo-inducing strangeness of the discovery. Everything we thought we knew about cosmic evolution had been turned upside down.

The implications were staggering. If the universe was accelerating, something was overpowering gravity on cosmic scales—and that something filled every cubic centimeter of apparently empty space. The discovery would eventually earn the 2011 Nobel Prize in Physics, but first it had to survive the scrutiny of a scientific community that was deeply skeptical of such an extraordinary claim.

The Detective Work Continues

But the evidence didn't stop with distant supernovae. As astronomers peered deeper into space, mapping the intricate cosmic web of galaxies, they found another crucial clue. The large-scale structure of the universe—the patterns of galactic clusters and superclusters, the vast filaments of matter connecting them, the enormous voids nearly empty of galaxies—could only be explained if there was an additional component to the universe's energy budget.

Think of it like trying to solve a jigsaw puzzle where most of the pieces are invisible. You have pieces representing all the matter you can see—stars, planets, gas clouds glowing in the darkness. You have pieces representing dark matter, revealed through its gravitational effects on visible matter and light. But when you try to assemble the cosmic picture, there are still huge gaps. The pattern doesn't make sense unless there's something else, something that doesn't just fill the gaps but actively shapes the overall design.

Computer simulations that attempted to reproduce the observed cosmic web consistently failed unless they included a dominant component that caused accelerated expansion. Without this mysterious energy, the simulations produced universes that looked nothing like what telescopes revealed. The structures were too dense, too clustered, evolving too quickly. Only when cosmologists added dark energy to their models did the simulated universes begin to resemble reality.

The cosmic microwave background—the afterglow of the Big Bang itself—provided the third crucial piece of evidence. This primordial light, carrying information from when the universe was only 380,000 years old, bore the signatures of a cosmos destined to be dominated by an unseen energy. The intricate patterns in this ancient radiation, mapped with exquisite precision by satellites like WMAP and Planck, could only be explained if dark energy made up about 68% of the universe's total energy content.

The angular size of hot and cold spots in the cosmic microwave background depended on the geometry of spacetime itself, which in turn depended on the total energy density of the universe. The observations revealed a universe that was geometrically flat—balanced precisely on the knife's edge between eternal expansion and eventual collapse. This flatness required a very specific energy budget: about 5% ordinary matter, 27% dark matter, and 68% dark energy.

Like ancient cartographers mapping uncharted realms, astronomers began assembling a mosaic of evidence—each observation a starburst of revelation outlining the invisible architecture of the cosmos. The combined evidence from supernovae, galactic surveys, and the cosmic microwave background painted a consistent picture: the universe was filled with an invisible energy that was pushing space itself apart. And this wasn't just any ordinary energy—it was energy that seemed to defy our basic understanding of how energy behaves.

The Paradox of Nothingness

The most bewildering aspect of dark energy was its relationship to emptiness itself. In our everyday experience, empty space is just that—empty. A void. Nothingness. When you remove all the air from a container, creating a vacuum, you expect to find... nothing. But quantum mechanics had long suggested otherwise, and dark energy forced us to confront the radical implications of quantum theory on cosmic scales.

Even the vacuum which is seemingly still and silent is a cauldron of invisible unrest, a simmering quantum sea where silence masks a perpetual storm of creation and annihilation.

Virtual particles pop into existence for the briefest instants before vanishing back into the quantum foam. This isn't science fiction—it's measurable reality with technological applications. The Casimir effect demonstrates that metal plates placed close together in vacuum are pushed together by the very nothingness between them, a phenomenon now exploited in ultra-sensitive micro-mechanical devices.

Picture a pot of boiling water. The surface is never still; bubbles are constantly forming and popping, each one briefly disturbing the smooth surface before disappearing. The quantum vacuum is like this, but instead of water and bubbles, it's space and particles, constantly bubbling with activity at the tiniest scales. These "bubbles" are virtual particles—pairs of particles and antiparticles that flash into existence and annihilate each other in timeframes so brief they make a nanosecond look like an eternity.

But these aren't just mathematical abstractions. Heisenberg's uncertainty principle allows these virtual particles to "borrow" energy from the vacuum, provided they pay it back quickly enough. The more energy they borrow, the faster they must repay the debt. For the most energetic virtual particles, the repayment period is so brief that they exist for less than 10^{-23} seconds—a timeframe so short that light can travel only a tiny fraction of the diameter of a proton.

During their fleeting existence, these virtual particles can be any type: electrons and positrons, quarks and antiquarks, photons materializing from pure vacuum energy. It's a never-ending dance of creation and destruction, happening everywhere, all the time, at scales so small we can't directly observe them. Yet their cumulative effects are very real and measurable.

The quantum foam gives rise to what physicists call zero-point energy—the lowest possible energy state of any quantum system. Even at absolute zero temperature, when all thermal motion stops, quantum fields continue to fluctuate. These fluctuations are not just theoretical curiosities; they have observable consequences ranging from the precise energy levels of atoms to the behavior of superconductors.

But here's where the cosmic accounting became truly bewildering. When physicists calculated how much energy this quantum activity should contribute to the expansion of the universe, they arrived at a number so large it was almost meaningless—a 1 followed by 120 zeros times larger than what we actually observe.

To grasp the enormity of this discrepancy, imagine you're trying to estimate the weight of a feather. You carefully measure its size, account for air density, consider the structure of the barbs and barbules. You do everything right according to your best theoretical understanding. But when scientists actually weigh the feather, they find your estimate was off by an almost unimaginable amount.

If your estimate for the feather's weight was off by this much, and the feather actually weighed one gram, your calculation would have suggested it weighed more than all the water in all the oceans on Earth, combined with the mass of Earth itself, multiplied by the mass of our entire Milky Way galaxy, and then multiplied again by the mass of all the galaxies in the observable universe. And even this incomprehensibly large mass would still be vastly smaller than your theoretical prediction.

In fact, if the actual weight of the feather was represented by the thickness of a human hair, your estimate would suggest a thickness stretching from Earth to the edge of the observable universe—and then millions of times beyond that! This is the magnitude of the discrepancy physicists face when calculating the energy of empty space compared to what we actually observe.

If their calculations were correct, the universe should be expanding so violently that atoms themselves couldn't hold together. Space would expand at breakneck pace, doubling in size every tiny fraction of a second. The temperature would drop so rapidly that the cosmic microwave background would cool from its current 2.7 Kelvin to near absolute zero in the blink of an eye. Any nascent structures—galaxies, stars, planets—would be ripped apart almost instantly. The resulting universe would be utterly inhospitable, a vast, empty expanse expanding so rapidly that not even atoms could hold together.

Yet here we were, embedded in a cosmos where dark energy provided just the gentlest cosmic acceleration—like the difference between a feather falling and a rocket ship blasting off. The universe was accelerating, but gently enough that galaxies had billions of years to form, stars had time to forge heavy elements, and life had the opportunity to emerge on at least one small rocky world.

The Paradox of Growing Emptiness

There was another puzzle hidden within the dark energy mystery, one that challenged our most basic intuitions about energy and conservation. Unlike the matter and radiation that fill the universe, dark energy didn't dilute as space expanded. Instead, it maintained a constant density—a fixed amount of energy in each cubic centimeter of space, even as the universe grew vaster and more spacious over time.

This constancy was one of the most puzzling aspects of dark energy. Normally, when something expands, its density decreases. If you have a box of gas and you double the volume of the box, the gas density becomes half what it was before. If you have a container of radiation and the container expands, the radiation becomes more diffuse, its energy density dropping as the photons spread out over a larger volume.

But dark energy doesn't follow this intuitive rule. As space expands, creating more volume, the total amount of dark energy increases proportionally. It's as if the universe has a built-in mechanism for generating more dark energy as it grows.

Picture the universe as a vast loaf of rising dough studded with raisins representing galaxies. As the dough rises, the raisins move farther apart—that's cosmic expansion. But here's the curious part: each cubic inch of dough comes with a rich, invisible flavor that represents dark energy. As the dough rises, creating more volume, the total flavor increases—not because it's being added from outside, but because more space means more flavored dough. The concentration of flavor stays exactly the same, but the total amount grows with the volume.

This peculiar trait of dark energy did more than surprise—it upended our most cherished assumptions about how energy behaves in an ever-stretching cosmos.

In our everyday experience, energy can neither be created nor destroyed—it can only change forms. But in an expanding universe governed by general relativity, the situation is more complex. Energy conservation, as we understand it in everyday physics, applies only within fixed reference frames. When spacetime itself is expanding, the global concept of energy conservation becomes ambiguous.

The universe doesn't violate conservation of energy by creating dark energy as space expands—rather, it reveals that our intuitive understanding of energy conservation doesn't apply to the universe as a whole. There's no cosmic bank account keeping track of total energy, because in general relativity, total energy isn't even a well-defined concept for the entire universe.

And so it is with dark energy: not created from some external reservoir, *but continually emerging from the fabric of being itself.* As the universe expands, creating new space, it simultaneously creates new dark energy to fill that space at the same density. The total amount increases, but the density remains constant—a property that would drive the universe's ultimate fate.

This suggested that dark energy wasn't just stuff that filled the universe, but a property of space itself. Every cubic meter of space contained about 6×10^{-10} joules of dark energy—an incredibly small amount (for comparison, a single grain of sand contains around 10^4 joules of chemical energy). To put this in perspective, if the energy density of dark energy were equivalent to having ten dollars, the energy in a grain of sand would be equivalent to having a hundred billion dollars.

Yet despite its minuscule density, dark energy played a dominant role in cosmic evolution because of the sheer vastness of space. When you sum up the dark energy across the entire observable universe—around 10^{86} cubic meters—it adds up to enough energy to drive the accelerated expansion of the entire cosmos. It's a case where the tiniest energy density, multiplied by the largest possible volume, yields the most significant force in the universe.

Einstein's Second Thoughts

As the evidence for dark energy mounted, scientists found themselves confronting a profound irony. The leading explanation for this mysterious force came from a concept that Einstein himself had introduced and later rejected as his "greatest blunder." The story of this conceptual resurrection illuminates how even our greatest thinkers sometimes abandon ideas before their time has come.

In 1917, Einstein was trying to construct a model of a static universe—one that would neither expand nor contract under the influence of gravity. This wasn't just a theoretical exercise; it reflected the scientific consensus of the time. The universe was thought to be eternal and unchanging, a vast stellar system that had always existed in roughly its current form.

To achieve this cosmic stasis, Einstein introduced a term into his field equations called the cosmological constant, represented by the Greek letter Lambda (Λ). This term represented a kind of inherent energy of space itself, a cosmic "spring tension" that could balance the attractive force of gravity. Just as you might add a repulsive force to balance an attractive one in a mechanical system, Einstein's cosmological constant provided the cosmic repulsion needed to keep the universe from collapsing under its own gravity.

The cosmological constant was mathematically elegant and theoretically consistent. It modified Einstein's equations in the simplest possible way, adding just one term that was the same everywhere in space and time. But Einstein introduced it reluctantly, viewing it as an arbitrary addition needed to force his equations to match the observed static universe.

But Einstein's static universe was not to be. In the 1920s, Edwin Hubble's observations of distant galaxies revealed something extraordinary: the universe was expanding. Galaxies were receding from us at velocities proportional to their distance—nearby galaxies moving slowly, distant ones racing away at tremendous speeds. The static universe was an illusion; we lived in a dynamic cosmos that was growing larger over time.

Faced with this observational evidence, Einstein abandoned his cosmological constant, reportedly calling it his greatest mistake. If the universe was expanding naturally under the influence of gravity and initial conditions from the Big Bang, there was no need for an additional repulsive force. The cosmological constant seemed like an unnecessary complication, a theoretical crutch that could be discarded now that observations had revealed the universe's true dynamic nature.

For decades, the cosmological constant languished in physics textbooks as a historical footnote—an example of how even genius can be led astray by incomplete observations. Theoretical physicists, following Einstein's lead, generally set Lambda equal to zero in their equations and focused on understanding the expansion history of a universe dominated by matter and radiation.

Yet like a buried archetype whose time had come again, Lambda rose from the ashes of abandonment. As evidence for cosmic acceleration grew in the late 1990s, the cosmological constant took on new life. No longer a mathematical trick to create a static universe, it became a possible explanation for the mysterious dark energy driving cosmic acceleration.

In this new context, the cosmological constant represented the inherent energy of the vacuum itself. Even in the absence of matter and radiation, space had an intrinsic energy density—a kind of cosmic ground state that remained constant as the universe expanded. As the universe grew, creating more space, it simultaneously created more of this vacuum energy. This constant supply of new energy was what drove the accelerated expansion.

The cosmological constant was elegant in its simplicity. It required no new particles, no exotic fields, no modifications to known physics. Space itself had energy, and that energy had a repulsive gravitational effect. In Einstein's general relativity, energy and momentum curve spacetime, and the cosmological constant represented a special kind of energy that curved spacetime in a way that caused acceleration rather than deceleration.

But this elegance came with a price—the cosmological constant problem that had haunted physics for decades. If the cosmological constant represented vacuum energy, then its value should be calculable from quantum field theory. And that calculation yielded the absurdly large number we encountered earlier—10^{120} times larger than observations suggested.

The Ancient Guide Appears

Once again, in my darkest hour of confusion, ancient wisdom offered an unexpected key. This time it came from Kashmir Shaivism, a mystical tradition that emerged in medieval India and described reality in terms that seemed to perfectly mirror what modern cosmology was discovering.

In this tradition, ultimate reality was understood as the eternal dance of Shiva and Shakti—pure consciousness and creative energy. But this wasn't mythology in the conventional sense; it was a sophisticated philosophical system that provided precise descriptions of how reality manifests from unmanifest potential.

Shiva represented the unmanifest potential, the infinite stillness from which all possibilities emerge. He was described as the silent witness, the unchanging awareness that observes all phenomena without being affected by them. Like the quantum vacuum in its ground state, Shiva contained infinite potential while appearing perfectly still and empty.

Shakti was the dynamic force that brought potential into manifestation, the creative power that transforms emptiness into the richness of existence. She was the cosmic energy that takes Shiva's infinite potential and expresses it as the world we experience. Like the virtual particles that emerge from quantum vacuum, Shakti represented the dynamic activity that arises from apparent stillness.

What captivated me wasn't the religious imagery, but the precise description of how these forces interacted. Kashmir Shaivism taught that Shiva and Shakti weren't separate entities but complementary aspects of one reality, forever dancing together through Spanda—the primordial vibration that gives birth to all phenomena.

Spanda was perhaps the most profound concept in this ancient system. It wasn't just a physical vibration; it was a principle that pervaded every level of reality. Spanda was described as the heartbeat of the universe, the whisper of the divine, the silent hum that permeates existence. It was the invisible thread that weaves the cosmic tapestry, the underlying rhythm that orchestrates the dance of creation, sustenance, and dissolution.

This ancient concept bore striking resemblance to quantum field theory's description of the vacuum. Just as Spanda was the fundamental vibration underlying all reality, quantum fields were fundamental vibrations from which all particles emerged. Virtual particles weren't separate entities but vibrations of underlying fields—just as all manifestation in Kashmir Shaivism emerged from Spanda's primordial vibration.

This wasn't mysticism foretelling physics. It was an ancient description of reality's structure that provided the conceptual framework I needed to understand dark energy's impossible nature.

The Dance of Emptiness and Form

In Kashmir Shaivism, the cosmic dance of Shiva and Shakti explained how the unmanifest becomes manifest without ever truly becoming separate from its source. Shiva's stillness contains infinite potential while Shakti's dynamism expresses that potential as the world we experience. Their union through Spanda—cosmic vibration—creates reality while preserving the underlying unity.

The tradition taught that Shiva and Shakti are inseparable, two aspects of a single reality rather than separate entities that somehow interact. Shiva without Shakti would be pure potential without any means of expression—like a musician without an instrument. Shakti without Shiva would be blind activity without conscious direction—like an instrument without a musician. Together, they form the complete reality that encompasses both potential and manifestation, stillness and activity, consciousness and energy.

Just as Shiva and Shakti danced in eternal embrace, giving rise to the manifest universe without ever becoming truly separate, the quantum vacuum danced with spacetime geometry itself. The constant creation and annihilation of virtual particles in the quantum realm was the cosmic dance giving rise to the energy that drove the universal expansion of spacetime.

The enormous energy associated with virtual particles in quantum field theory could be seen as a reflection of Shakti's raw, primal power—the cosmic energy that drives the dance of creation. This energy was real and measurable, manifesting in phenomena like the Casimir effect and the anomalous magnetic moment of the electron. Yet somehow, when it came to driving cosmic expansion, this energy appeared to be tamed, regulated, reduced to the gentle influence we observe as dark energy.

But the key insight came from understanding Spanda—the primordial vibration that mediates between unmanifest and manifest. In our cosmic context, this vibration was the coupling between quantum fluctuations and spacetime geometry itself.

Could it be that spacetime itself might vibrate in response to quantum activity? And if so, how would the vibration of spacetime in turn affect the energy from quantum activity?

The Missing Piece

Traditional physics had been treating spacetime as a passive stage upon which quantum drama unfolds. Physicists calculated the energy of virtual particles while assuming that spacetime remained unchanging, like a theater that never responds to the performance taking place upon it.

This separation seemed natural and convenient. General relativity dealt with the large-scale structure of spacetime, while quantum mechanics dealt with the small-scale behavior of particles and fields. The two theories worked beautifully in their respective domains, so why complicate matters by considering their interaction?

But what if spacetime was an active participant in the quantum dance? What if the apparent emptiness of vacuum and the fabric of spacetime were locked in a perpetual embrace, each shaping and responding to the other?

Einstein's own equations suggested this should be the case. General relativity states that matter and energy curve spacetime, and spacetime curvature affects how matter and energy behave. If virtual particles carry energy and momentum—even briefly—they should create ripples in spacetime geometry. These ripples would, in turn, influence how quantum fields behave, potentially creating a feedback loop that could regulate the system.

This insight led me to a radical reinterpretation of the cosmological constant problem. The absurdly large energy density calculated for the quantum vacuum wasn't wrong—it was incomplete. It assumed that spacetime remained unmoved by quantum activity, when Einstein's own equations insisted that spacetime should respond to energy and momentum.

Even the fleeting virtual particles of quantum foam should cause ripples in the fabric of space and time itself.

This reciprocity hinted at a deeper unity: that the quantum vacuum and spacetime geometry are not separate domains, but harmonized expressions of a single, self-tuning reality.

The universe might have evolved mechanisms to prevent the vacuum energy from running away to infinity, not through arbitrary fine-tuning, but through the natural dynamics of quantum-gravitational interaction.

The Self-Regulating Cosmos

Following this logic, I developed a mathematical framework that treated quantum fluctuations and spacetime geometry as partners in a cosmic dance. Instead of quantum activity occurring against a fixed spacetime background, both domains would influence each other through what I called stochastic metric fluctuations—random variations in the geometry of spacetime itself. For those interested, my paper on this topic that is currently in the process of publishing is included in the Appendix.

My approach utilized the Einstein-Langevin equation—a version of Einstein's gravitational equations that includes random, stochastic influences. Just as Brownian motion describes how invisible molecular collisions cause visible particles to jitter randomly, this framework described how invisible quantum fluctuations cause spacetime itself to jitter at the smallest scales.

The idea wasn't entirely unprecedented. Physicists had earlier laid conceptual foundations on the general properties of stochastic gravity along with the mathematical tools required to express them. However, my approach specifically applied these to regularize vacuum energy and used a phenomenological framework to link it with observed values of vacuum energy to find the physical scale characterizing vacuum-spacetime coupling.

The mathematics revealed something remarkable. When quantum virtual particles flickered into existence, they would create tiny ripples in spacetime geometry according to Einstein's equations. These ripples were incredibly small—far smaller than could be detected by any conceivable experiment—but they were present in principle throughout the quantum vacuum.

These spacetime fluctuations would then influence how future virtual particles behaved. The coupling between quantum fields and curved spacetime was well understood from quantum field theory in curved spacetime, a mature branch of theoretical physics.

What was new was considering the feedback effect: *how quantum activity could influence spacetime geometry, which in turn would influence quantum activity.*

The Universe's Hidden Score

The implications extended far beyond solving a numerical puzzle. My framework suggested that dark energy wasn't some mysterious substance added to the universe, but the natural consequence of vacuum fluctuations coupling to spacetime itself. It reframed quantum mechanics and general relativity not as rivals vying for dominance, but as harmonics in a deeper symphony—complementary lenses through which one unified reality reveals itself.

Every cubic centimeter of space contained the same amount of this vacuum-spacetime coupling energy, explaining why dark energy appeared uniform throughout the cosmos. Unlike matter that clumped into galaxies and stars due to gravitational attraction, or radiation that clustered around matter sources, this coupling energy was a property of space itself. It couldn't clump because it was woven into the very fabric of spacetime.

The uniformity of dark energy also explained one of its most puzzling aspects: why it maintained exactly the same density everywhere we looked. Whether astronomers observed nearby galaxies or the most distant quasars at the edge of the observable universe, they found the same evidence for the same dark energy density. This perfect uniformity had puzzled cosmologists because most cosmic phenomena showed at least some variation from place to place.

But if dark energy emerged from the coupling between vacuum fluctuations and spacetime geometry, its uniformity was inevitable. Quantum field theory predicted that vacuum fluctuations had the same statistical properties everywhere in space, and general relativity described spacetime as having the same geometric properties everywhere on large scales. Their coupling would naturally produce the same energy density throughout the cosmos.

As the universe expanded, creating new space, it also created new coupling energy—not violating conservation laws, but revealing that energy conservation works differently in an expanding universe than in our everyday experience. In Einstein's general relativity, there's no universal conservation law for total energy when spacetime itself is changing. The cosmic energy budget isn't fixed like money in a bank account—it's more like the total wetness of water, which isn't conserved when you create more water.

This was why dark energy maintained constant density even as space expanded. It wasn't that energy was being created from nothing, but that the very fabric of spacetime carried this energy as an intrinsic property, like the wetness of water or the transparency of glass. You can't separate water from its wetness, and you can't separate spacetime from its intrinsic vacuum-geometry coupling energy.

Beyond the Laboratory

The deeper revelation was philosophical as much as physical. If vacuum fluctuations and spacetime geometry were locked in an eternal dance, then the apparent emptiness of space was actually the most energetically active aspect of reality. What we perceived as nothingness was actually everything—the creative source from which all phenomena emerge.

This insight resonated with the deeper current flowing through this work—the Unified Conscious Matrix—where matter, mind, and vacuum weave a single tapestry of creative emergence.

Just as consciousness and information were fundamental rather than emergent, the energy of empty space was fundamental rather than incidental. We weren't living in a universe where consciousness emerged from dead matter floating in empty space. We were living in a universe where consciousness, information, and the creative energy of vacuum were the primary constituents from which matter and space themselves emerged.

Every atom in your body was held together by forces that emerged from quantum field fluctuations. The electromagnetic force that binds electrons to nuclei, the strong force that holds protons and neutrons together, even the weak force responsible for radioactive decay—all of these fundamental interactions arose from the activity of quantum fields that filled apparently empty space.

Every chemical bond that made biological molecules possible, every electrical impulse that carried information through your nervous system, every thought that flickered through your consciousness—all of it depended on the underlying quantum activity that filled the vacuum between particles. The space between your thoughts was as alive with creative potential as the vacuum between distant galaxies.

This perspective offered more than just a solution to the cosmological constant problem—it illuminated a path toward understanding reality at its most fundamental level. Perhaps our quest for a unified "theory of everything" had been hampered by our insistence on treating spacetime and quantum phenomena as separate domains. The Unified Conscious Matrix suggested that they were, in fact, complementary expressions of a single, unified whole.

The framework also suggested new ways of thinking about Consciousness itself. If vacuum fluctuations could influence spacetime geometry, and spacetime geometry could influence quantum fields, then perhaps Consciousness—which seemed to play a fundamental role in quantum measurement—might also couple to spacetime in ways we were only beginning to understand.

The Dance Continues

Standing in our cosmic forest with this new understanding, I realized we had uncovered the fourth great truth of the Unified Conscious Matrix: even apparent emptiness is alive with creative potential. The vacuum that fills space isn't nothing—it's the most energetic thing in the universe, engaged in an eternal dance with spacetime geometry that drives cosmic evolution itself.

The ancient seers had intuited what our equations now reveal—that the void is not a lack, but a latent plenitude, a womb of infinite becoming.

They understood that emptiness and fullness, silence and sound, stillness and activity were complementary aspects of one creative reality. Modern physics was rediscovering it through experiment and equation, finding mathematical descriptions of what mystics had long understood.

In Kashmir Shaivism, this dance was called the union of Shiva and Shakti through Spanda—the primordial vibration that gives birth to all phenomena. In the physics of my Unified Conscious Matrix, it was the coupling of vacuum fluctuations to spacetime geometry that gave rise to the dark energy driving cosmic acceleration. Both traditions pointed to the same fundamental truth: reality is not built from separate components but emerges from the creative interaction of consciousness, energy, and space itself.

The parallels went even deeper. Just as Spanda was described as the heartbeat of the universe, the coupling between vacuum fluctuations and spacetime created a kind of cosmic rhythm—a fundamental frequency that characterized how quantum activity and geometric curvature influenced each other. In the equations of my framework, this rhythm resonates at 1.8 THz, a frequency that would soon be accessible to experimental investigation.

The Living Void

This revelation transformed my understanding of existence at the most fundamental level. We weren't isolated beings floating in dead empty space. We were expressions of a universe where even the vacuum was alive with creative energy, where consciousness and information and vacuum fluctuations were all aspects of one unified field of awareness and potential.

Whether between your thoughts or between the stars, that same primordial pulse—Spanda to the mystic, vacuum-spacetime coupling to the physicist—permeates all gaps, animating every silence with hidden music.

Your very existence depended on this creative dance of emptiness and form. The carbon atoms in your DNA had been forged in stellar cores powered by quantum processes. The electrical activity in your brain relied on electromagnetic fields that emerged from vacuum fluctuations. Your consciousness itself might be a localized expression of the same awareness that guided quantum measurement and the emergence of classical reality from quantum potential.

The Cosmic Symphony

As we prepare to explore even deeper mysteries in the chapters ahead, we carry this transformative understanding: the universe is not built from matter floating in empty space, but from the creative interaction of consciousness, information, and the energetic vacuum that gives birth to both space and matter.

Dark energy is no foreign agent—it is the native rhythm of reality itself, the vacuum's creative whisper entwined with the pliant curvature of spacetime.

We live in a universe where emptiness itself is full, where nothingness vibrates with infinite possibility, where the apparent void between all things is actually the source from which all things emerge.

The ancient dance of Shiva and Shakti continues in every cubic centimeter of space, in every quantum fluctuation, in every moment of cosmic expansion. The primordial vibration called Spanda resonates, if my theory holds up in experiments, at 1.8 THz, the frequency where vacuum fluctuations and spacetime geometry achieve their most intimate coupling. And we are not separate observers of this dance—we are the universe awakening to its own creative nature through forms of consciousness that emerge from the same unified field.

Regardless of what those experiments reveal, the search itself has already transformed our understanding. We've moved from seeing dark energy as an alien force disrupting cosmic harmony to recognizing it as the very source of that harmony—the rhythm section of reality's grand symphony, keeping time for the cosmic dance of consciousness and creation that plays out across all scales, from quantum to cosmic, from virtual particles to living beings awakening to their role in the infinite creative potential of existence itself.

In the end, the mystery of dark energy has led us to a profound truth: we are not separate from the vacuum that surrounds us, not isolated from the emptiness between galaxies, not disconnected from the creative forces that drive cosmic evolution.

We are expressions of the vacuum's own creativity, manifestations of emptiness awakening to its own infinite fullness, participants in a cosmic dance where consciousness, information, energy, and space itself are all aspects of one magnificent, unified, ever-creative reality.

CHAPTER FIVE

The Primal State of Everything

"The Universe emerged from an infinite, undifferentiated potential, a single source from which all diversity and forms arose"

—Ancient wisdom from the Atharvaveda tradition of India, written 3,300 years ago

The discoveries we had made—that reality was unified, that consciousness and information were fundamental, that everything was interconnected—had shattered my understanding of the universe. But they also raised the ultimate question: if this cosmic dance of consciousness and information created everything we experience, what was the source of the dance itself?

The answer would lead me to the most profound revelation yet: that everything we call real emerges from a state so primordial, so utterly beyond our ordinary experience, that ancient sages could only describe it as "that from which all things arise and into which all things return."

The Cosmic Tapestry Unraveled

Standing in our cosmic forest, I had discovered that what appeared to be two separate realms—the visible SpacetimeFabric and the hidden QuantumLoom—were actually interwoven aspects of one unified reality. The visible trees above ground and the vast root networks below weren't just connected; they were expressions of one living system appearing as two complementary domains.

But now a deeper question haunted me: which came first? Was spacetime the fundamental stage upon which quantum dramas unfolded, or was there something more primordial still? It was as if what we had taken to be the forest's bedrock turned out to be a shallow crust—beneath which stretched fathomless strata of primordial reality.

The clues had been hiding in plain sight throughout our journey. Every quantum mystery we'd encountered—wave-particle duality, entanglement, the delayed choice experiments—pointed toward the same conclusion: spacetime itself might not be fundamental. The solid stage upon which the cosmic drama appeared to unfold might itself be a projection, an emergent property of something far more basic.

Think of a movie theater. The audience sees vivid three-dimensional action on the screen, complete with depth, movement, and drama. But the projectionist knows that all this apparent reality emerges from patterns of light and shadow cast by a two-dimensional film. The "reality" on screen is real enough to the audience, but it's actually a projection from something more fundamental.

What if our entire spacetime universe was like that movie—a projection from some unimaginably more fundamental domain?

The evidence was overwhelming once I learned to see it. Quantum particles could exist in multiple places simultaneously until measured—how could this be possible if space was truly fundamental? If space were the bedrock of reality, particles would have to be somewhere specific at all times. The fact that they could exist in superposition—everywhere and nowhere until observed—suggested that space itself emerged from their observation.

Entangled particles remained connected across vast distances instantly—how could this happen if spacetime's structure was absolute? If space were fundamental, nothing could influence anything else faster than light could travel between them. Yet entanglement demonstrated correlations that transcended spatial separation entirely, as if distance itself were somehow secondary to a deeper level of connection.

Information could apparently flow backward in time in delayed choice experiments—how could this occur if temporal sequence was basic to reality's architecture? If time were fundamental, the past would be fixed and unreachable from the future. Yet quantum mechanics showed us events where future choices seemed to retroactively determine past outcomes.

Each paradox whispered the same revolutionary truth: the QuantumLoom wasn't just hidden beneath spacetime—it was the primordial state from which spacetime itself emerged.

The Quantum Womb of Reality

In the double-slit experiment, electrons existed in superposition until measured—they literally occupied no definite position in space until spacetime "crystallized" around them through observation.

This wasn't just about measurement changing outcomes. It was about measurement creating the very stage upon which outcomes could occur.

Consider the implications carefully: if particles don't have definite positions until measured, and if position is what we mean by "location in space," then space itself might not exist until the act of measurement calls it into being. The quantum realm—what I call the QuantumLoom—operates in a domain where our concepts of "here" and "there," "before" and "after," simply don't apply.

This sounds impossible until you realize that we've been thinking backwards. We assume space and time are the fundamental containers in which events occur. But what if it is not space and time that cradle events, but events themselves that give rise to space and time?

What if measurement doesn't just reveal where a particle is, but participates in creating the spatial framework that gives meaning to "where"?

The mathematics supported this stunning possibility. In quantum field theory, the most successful theory in physics, particles aren't fundamental entities moving through space. They're excitations of underlying fields that exist in abstract mathematical spaces with no direct spatial meaning. The Dirac equation describes electrons as excitations in a field that spans all of space, but the "space" it describes is mathematical, not physical.

What we experience as particles in definite locations are temporary crystallizations of these field excitations into the familiar three-dimensional space we call home. It's like watching waves on an ocean—the waves appear to be discrete entities moving across the water's surface, but they're actually temporary patterns in the continuous medium of the ocean itself.

But if space emerges from something deeper, what about time? Here the evidence was even more compelling. The delayed choice quantum eraser experiments we'd explored showed that future measurements could retroactively determine past events. This impossibility became comprehensible only if we recognized that the QuantumLoom operates outside temporal sequence entirely.

In relativity, we learned that time is relative to the observer's frame of reference. But quantum mechanics suggested something far more radical: that time itself might be an emergent property—a kind of crystallization of deeper, atemporal relationships in the quantum domain.

Consider how this revolutionizes our understanding of causality. In classical physics, cause precedes effect in a clear temporal sequence. But in the quantum domain, all potential causes and effects might exist simultaneously in a state of superposition, with measurement crystallizing specific causal chains into the temporal sequence we experience.

Time, like space, might be an emergent property—not the fundamental framework within which events occur, but the structure that emerges when quantum possibilities crystallize into classical actualities.

When Distance Becomes an Illusion

Quantum entanglement provided the most direct evidence for the primacy of the QuantumLoom. When two particles become entangled, they don't just share mysterious correlations across space—they reveal that spatial separation itself is somehow secondary to their deeper connection.

Einstein called this "spooky action at a distance," but his phrase revealed the limitation of thinking in terms of spatial distance. In the QuantumLoom, entangled particles aren't separate entities that somehow communicate across space. They're aspects of a single reality that was never divided in the first place.

Think of a hologram. When you break a holographic plate into pieces, each fragment contains the entire image, not just a part of it. The "separation" of the fragments is real at the level of the physical plate, but the holographic image they encode remains unified. You can take the fragments to opposite sides of the galaxy, but the image each contains remains complete and identical.

Similarly, entangled particles may appear separate in spacetime, but in the QuantumLoom—the deeper level where their wave function exists—they remain unified. The "distance" between them in spacetime is like the distance between holographic fragments—real at one level, but irrelevant at the level where their fundamental unity is maintained.

This perspective transformed my understanding of Bell's theorem and the experimental violations of Bell inequalities. Alain Aspect's groundbreaking experiments didn't just prove that quantum mechanics was right and local realism was wrong—they demonstrated that reality operates according to principles that transcend spatial locality altogether.

Entanglement does not signal across distance—it discloses a deeper unity, a realm where the illusion of space has not yet hardened into form. When we measure one particle, we're not sending a signal to its partner—we're accessing a unified quantum state that encompasses both particles simultaneously.

This explained why entanglement doesn't violate relativity's prohibition on faster-than-light communication. No information travels between the particles because they were never truly separate in the domain where their correlation is established—the primordial QuantumLoom.

The mathematical description of entanglement supports this interpretation. Entangled particles are described by a joint wave function that cannot be factored into separate parts. This isn't just a mathematical convenience—it reflects a physical reality where the particles exist as an irreducible unity in the quantum domain, even as they appear separate in spacetime.

Consider the broader implications: if entanglement reveals the secondary nature of spatial separation, what does this say about all the apparent separations in our world? The distance between atoms in your body, the space between Earth and distant stars, the gulf between your consciousness and the objects you perceive—all of these separations might be features of spacetime's crystallized structure rather than fundamental aspects of reality.

The Universe's Birth Certificate

Cosmology provided the most spectacular confirmation of the QuantumLoom's primordial nature. The cosmic microwave background—the afterglow of the Big Bang that fills all of space—carries the fingerprints of quantum processes that operated before space and time as we know them existed.

This ancient light, released when the universe was only 380,000 years old, bears the imprint of even earlier events. The nearly scale-invariant spectrum of fluctuations in this cosmic radiation revealed something profound: the seeds of cosmic structure originated as quantum fluctuations that were stretched to cosmic scales during the inflationary epoch.

But here's the crucial insight: these weren't fluctuations within a preexisting spacetime framework—they were fluctuations of the QuantumLoom itself that became spacetime as they crystallized.

Think of it this way: imagine a vast ocean of possibility suddenly freezing into ice. The ice crystals that form carry the pattern of the currents and waves that existed in the liquid state. Similarly, the structure of our universe carries the pattern of quantum fluctuations that existed in the QuantumLoom before spacetime crystallized.

The statistical properties of these cosmic seeds were precisely what quantum field theory predicted for fluctuations in a pre-geometric quantum domain. Their Gaussian distribution—the familiar bell curve pattern—was exactly what would result from quantum processes. Their scale-invariance, meaning they looked the same at all size scales, was a signature of quantum origin rather than classical processes.

Most remarkably, these fluctuations displayed the kind of subtle correlations that quantum mechanics predicted but classical physics could not explain. When cosmologists analyzed the cosmic microwave background with exquisite precision, they found patterns that could only have originated from quantum entanglement operating on cosmic scales during the primordial epoch.

But the most stunning revelation came from understanding what cosmic inflation actually represented. During inflation, the universe expanded faster than light—something impossible within the established spacetime framework. This paradox dissolved when I realized that inflation wasn't expansion of space within a preexisting framework, but the crystallization of space itself from the QuantumLoom.

Envision a cloud of quantum potential condensing in an instant—like unseen vapor crystallizing into patterned ice, the invisible made manifest.

The "expansion" wasn't movement through space but the very creation of space as quantum potentiality crystallized into geometric reality.

This perspective resolved several long-standing cosmological puzzles. The flatness problem—why space is so precisely flat on cosmic scales—found natural explanation. In the QuantumLoom, before geometric concepts applied, there was no meaningful distinction between flat and curved. Flatness emerged naturally as the simplest geometric state that could crystallize from pure quantum potentiality.

The horizon problem—why distant regions of the universe that could never have communicated appear to have the same temperature—also dissolved. In the QuantumLoom, before space and time had crystallized, "distant regions" didn't exist as separate entities. The uniformity we observe reflects the fundamental unity of the primordial state before spatial separation emerged.

Matter's Quantum Secret

Einstein's famous equation $E=mc^2$ took on a new meaning in this context. The interchangeability of matter and energy wasn't just a curious property of substances within spacetime—it was evidence that both matter and energy were crystallizations of something more fundamental.

In quantum field theory, particles are excitations of underlying fields. An electron isn't a tiny ball orbiting an atomic nucleus—it's a localized excitation in the electron field that spans all of space. A photon isn't a little packet of light—it's an excitation in the electromagnetic field. But fields, I realized, were themselves crystallizations of the QuantumLoom into the geometric framework of spacetime.

What we call "particles" are like standing waves in this crystallized structure—temporary stable patterns in the ongoing dance between the QuantumLoom and its spacetime manifestation. They're not things moving through space so much as stable patterns in the way space itself is organized.

The Casimir effect, where metal plates in vacuum are pushed together by "empty" space, wasn't just a demonstration of virtual particle pressure. It was direct evidence of the QuantumLoom's influence on spacetime, showing how the primordial quantum domain could exert measurable forces within the classical world.

Even more remarkably, this framework suggests a new understanding of dark matter—that mysterious substance that makes up 27% of the universe but has never been directly detected despite decades of searching.

What if dark matter wasn't some exotic new particle, but ordinary matter that remained primarily in the QuantumLoom state—matter that hadn't fully crystallized into definite spacetime locations?

The Schrödinger equation revealed a tantalizing clue: while unmeasured particles exist in probabilistic clouds of position and momentum, their mass remains definite. This suggested that mass might be a property that transcends the quantum-classical divide, remaining constant whether matter exists in the QuantumLoom or crystallizes into spacetime.

Consider the implications: if cosmic matter could exist in various degrees of crystallization—some fully manifest in spacetime, others remaining in quantum superposition—it would explain dark matter's gravitational effects without electromagnetic interaction. Quantum-state matter would still possess mass and thus curve spacetime gravitationally, but without definite positions, it couldn't engage in electromagnetic processes that require spatial localization.

I have developed these ideas into a technical paper that is in the process of being published. This model explains why dark matter interacts gravitationally but not electromagnetically with ordinary matter. It's not that dark matter particles are inherently different from ordinary matter—it's that they haven't undergone the measurement processes necessary to fully crystallize into spacetime locations.

The Holographic Revelation

The holographic principle provided another stunning confirmation of spacetime's emergent nature. This principle, arising from black hole physics, suggested that the information content of any volume of space could be encoded on its boundary—like a hologram storing three-dimensional information on a two-dimensional surface.

The discovery emerged from studying black hole entropy. Stephen Hawking had shown that black holes have temperature and entropy proportional to their surface area, not their volume. This was shocking because entropy usually measures the number of possible microscopic states of a system, and this number should depend on volume, not surface area.

But if space is holographic, what is the more fundamental reality from which it emerges? The answer pointed directly to the QuantumLoom—a domain of pure information and relationship that could encode the patterns that we experience as three-dimensional space.

This wasn't just a mathematical curiosity. It suggested that space itself was a kind of information processing—a cosmic computation running on the quantum substrate of the QuantumLoom. What we experience as three-dimensional reality may be but a luminous simulation—born from the deeper, silent computations of the quantum substrate.

Think of a video game. The characters in the game experience a three-dimensional world with objects, forces, and interactions. But underneath, it's all information processing—patterns of bits being manipulated according to computational rules. The holographic principle suggested that our spacetime might be similar: a three-dimensional "game world" arising from information processing in the QuantumLoom.

The holographic principle also explained why quantum entanglement seemed to "weave" spacetime together. If space emerges from quantum information processing, then entanglement—the fundamental quantum relationship—would indeed be the thread from which spatial geometry is woven.

Recent work in quantum gravity has shown that entanglement and spacetime geometry are intimately connected. The Einstein-Rosen bridges that connect different regions of spacetime in general relativity are mathematically equivalent to entanglement connections in quantum mechanics. Space, it seems, is literally woven from quantum relationships.

This perspective transformed Einstein's general relativity from a theory about how matter curves preexisting spacetime into a description of how matter and spacetime co-emerge from their common source in the QuantumLoom. The curvature of spacetime wasn't just response to matter—it was part of the ongoing crystallization process by which both space and matter maintained their interdependent existence.

Ancient Wisdom as Cosmic Map

Once again, ancient wisdom provided the conceptual framework I needed to comprehend these revolutionary insights. But this time, the parallel was so precise it took my breath away.

The *Atharvaveda*, composed over three millennia ago, described the universe emerging from "an infinite, undifferentiated potential, a single source from which all diversity and forms arose." This wasn't primitive cosmology—it was a precise description of what modern physics was discovering about the QuantumLoom.

The ancient text continued: "From non-being came being." This paradoxical statement perfectly captured the emergence of spacetime from the QuantumLoom. The "non-being" wasn't nothingness but rather a state of being so fundamental that ordinary categories of existence didn't apply to it—exactly what we'd discovered about the quantum domain.

But the most profound parallel came from Kashmir Shaivism's concept of Spanda—the primordial vibration that gives birth to all phenomena. According to this tradition, ultimate reality (Shiva) exists as pure consciousness in a state of dynamic potentiality. Through Spanda—cosmic vibration—this potential manifests as Shakti, the creative force that becomes the world of form.

This ancient description perfectly matched what I was discovering about the relationship between the QuantumLoom and spacetime. The QuantumLoom was like Shiva—pure potentiality beyond space and time. Spacetime was like Shakti—the creative manifestation of that potentiality. And the process of crystallization was like Spanda—the fundamental vibration that transforms potential into manifest reality.

The *Spanda Karika*, a foundational text of Kashmir Shaivism, described this cosmic vibration: "The divine creative pulsation (Spanda), which is the very essence of the Supreme Shiva, is constantly rising and falling. It is the source of the manifestation, maintenance, and reabsorption of the Universe."

This wasn't mere poetry—it was a precise map of how reality unfolds, drawn from direct experiential insight. The "rising and falling" of Spanda corresponded to the constant fluctuations of the quantum vacuum. The quantum field doesn't just sit still—it's in constant motion, with virtual particles constantly appearing and disappearing in a cosmic dance of creation and destruction.

The "manifestation, maintenance, and reabsorption" described the process by which quantum possibilities crystallized into spacetime reality, persisted as classical phenomena, and eventually dissolved back into the primordial state. Every particle that exists does so temporarily, eventually decaying or transforming back into energy, which itself dissolves back into the quantum vacuum from which it emerged.

The sages had perceived through contemplative practice what modern physics was discovering through mathematics: reality is not a collection of static objects but a dynamic process of continuous creation, maintenance, and dissolution—a cosmic dance of consciousness manifesting as the world of form.

The Cosmic Heartbeat

The concept of Spanda provided a crucial insight I had been missing: the relationship between the QuantumLoom and spacetime wasn't static but dynamic. Reality wasn't a one-time event—it was a continuous heartbeat, a rhythmic pulse where spacetime arises and dissolves into its quantum origin, moment by moment.

This dynamic perspective explained why quantum effects persisted even in the apparently classical world. Virtual particles, quantum tunneling, zero-point energy fluctuations—all were evidence of the QuantumLoom's ongoing influence, the primordial state continuously reasserting itself within the crystallized structure of spacetime.

Every particle interaction, every quantum measurement, every moment of observation was a note in this cosmic symphony. The apparent stability of the classical world was like the stability of a river—constant on the macroscopic scale but composed of countless individual water molecules in constant motion. Similarly, the stability of spacetime emerged from countless quantum processes, each one a moment in the ongoing dance between potential and actuality.

This perspective also suggested a new understanding of cosmic evolution. Rather than seeing the universe as a closed system winding down toward heat death according to the second law of thermodynamics, I began to envision it as an open system, continuously renewed by its connection to the inexhaustible creativity of the QuantumLoom.

This perspective resonated with ancient Vedantic concepts of cosmic cycles. The *Upanishads* described the universe as continuously emerging from, being sustained by, and eventually dissolving back into Brahman—the unchanging ground of being. This wasn't seen as a linear process with a beginning and end, but as a cyclical process without ultimate beginning or end.

Just as these texts envisioned a cyclical cosmos rather than a linear progression toward either eternal expansion or final collapse, the QuantumLoom model suggested that regions of spacetime might continually dissolve back into the quantum substrate, only to re-emerge in new forms. The universe wasn't dying but transforming, not running down but continuously renewing itself from its inexhaustible source.

The Arrow of Time Revealed

This dynamic understanding also illuminated one of physics' deepest mysteries: the arrow of time. Why does time flow from past to future when the fundamental laws of physics work equally well in either direction? Why do we remember the past but not the future? Why does entropy increase rather than decrease?

The QuantumLoom model suggested an elegant answer: the arrow of time wasn't built into the laws of physics but emerged from the process of crystallization itself. As quantum potential crystallized into classical actuality, it naturally created a directionality—a flow from the undifferentiated QuantumLoom toward the structured reality of spacetime.

Imagine water crystallizing into ice. The process naturally creates a direction—from fluid to structured state. The water molecules don't move from past to future in some absolute sense, but the crystallization process creates an intrinsic direction from less organized to more organized structure.

Similarly, the ongoing crystallization of the QuantumLoom into spacetime created the temporal direction we experience as the flow of time. Time wasn't just about clocks ticking or things aging—it was about the universe's fundamental creative process, the ongoing transformation of potential into actuality that defined cosmic evolution itself.

This perspective suggested that time wasn't just a dimension like space, but the signature of the creative process itself. Every moment was a moment of crystallization, a moment when new possibilities from the QuantumLoom manifested as spacetime reality.

It also explained why we experience time as irreversible. While individual physical processes might be time-reversible, the overall crystallization process had a built-in directionality. You could melt ice back into water, but the cosmic process of crystallization from quantum potential into classical reality provided the fundamental arrow that gave time its direction.

The increase in entropy—the universe's tendency toward greater disorder—could be understood as a natural consequence of this crystallization process. As quantum possibilities crystallized into classical realities, the number of possible states increased, leading to the statistical tendency toward greater disorder that we observe as the second law of thermodynamics.

But this wasn't a death sentence for the universe. The QuantumLoom represented unlimited potential for new crystallizations, new forms of organization, new expressions of creative possibility.

The Infinite Creative Source

Perhaps most profoundly, this understanding revealed the universe not as a finite system running down toward equilibrium, but as an expression of infinite creativity continuously manifesting through the crystallization process.

The QuantumLoom represented unlimited potential—not just the potential for different arrangements of matter and energy within spacetime, but the potential for spacetime itself in all its possible forms. Every possible universe, every conceivable arrangement of space and time, every potential physical law—all existed as possibilities within the primordial quantum state.

Our universe was just one crystallization from this infinite sea of potential, one particular pattern that had emerged from the cosmic superposition. But the source remained inexhaustible, continuously capable of manifesting new realities, new forms of space and time, new expressions of the fundamental creative principle.

This vision echoed the timeless recognition that Consciousness is not passive but intrinsically, endlessly creative.

Just as individual consciousness continuously creates the stream of thoughts, perceptions, and experiences that constitute personal reality, cosmic Consciousness—the fundamental awareness of the QuantumLoom—continuously created the stream of quantum crystallizations that constituted physical reality.

Every thought you think, every choice you make, every moment of awareness participates in this cosmic creativity. You're not separate from the process but an expression of it—a localized way the universe creates and experiences itself.

The Living Universe

Standing at the threshold of this understanding, I realized we had uncovered not just a new theory about cosmic origins, but a new vision of reality itself. The universe wasn't a collection of objects moving through empty space—it was a living process, the ongoing creative expression of consciousness manifesting through the continuous crystallization of quantum potential into spacetime actuality.

We were never outsiders looking in—we were the cosmos gazing at itself through the lens of living awareness.

Our consciousness was locally focused but universally grounded, individual waves in the infinite ocean of awareness that was the QuantumLoom itself.

This understanding transformed every aspect of existence. The space between atoms in our bodies was alive with quantum potential. The thoughts arising in our minds were local crystallizations of the same creative process that formed galaxies and stars. The consciousness through which we experienced reality was intimately connected to the fundamental Consciousness that guided cosmic evolution at every scale.

Even our sense of being separate individuals was a kind of crystallization—a way the universal Consciousness focused itself through particular biological structures while never losing its essential unity. We were like whirlpools in a stream, appearing separate and distinct while actually being temporary patterns in the continuous flow of awareness.

This vision offered profound hope and meaning. If the universe was fundamentally creative consciousness expressing itself through infinite forms, then consciousness, creativity, and meaning weren't rare accidents but the basic nature of reality itself. We lived in a cosmos that was not only conscious but infinitely creative, continuously exploring new possibilities for existence, awareness, and experience.

The Fifth Revelation

We had uncovered the fifth great truth of the Unified Conscious Matrix: the QuantumLoom is the primal state from which everything emerges. Spacetime, matter, energy, and even the classical laws of physics are crystallizations of a more fundamental quantum reality that operates beyond space and time.

This wasn't just a new scientific theory—it was a return to the primordial understanding that Consciousness, not matter, is the foundation of reality. The ancient sages had known this truth through direct contemplative insight. Modern physics was rediscovering it through mathematics and experimentation. And in their convergence lay a revolutionary understanding of our true nature and place in the cosmos.

In recognizing the QuantumLoom as the primal state, we don't just gain a new cosmology—we remember our own deepest nature. We are not matter that accidentally gained awareness—we are Consciousness itself, briefly localized through form, forever rooted in the infinite source from which all arises.

The cosmic dance continues, but now we know our role in it: we are not separate from the dance—we are the dance awakening to its own infinite nature through countless forms of awareness, each one a unique expression of the one Consciousness that is the source, sustenance, and destination of all that exists.

As we prepare for the next phase of our journey, we carry this transformative understanding: we live in a universe that is conscious, creative, and infinite in its potential. Every moment offers new possibilities for the crystallization of quantum potential into spacetime actuality. Every choice we make participates in the ongoing creation of reality itself.

In the next chapter, we'll explore how this primordial Consciousness expresses itself through the magnificent complexity of biological life, weaving the patterns of DNA and consciousness into the tapestry of living systems that bridge the quantum and classical domains in ways that continue to revolutionize our understanding of what it means to be alive in a Conscious universe.

CHAPTER SIX

Life - The Quantum Self Awakening

"The most challenging frontier in biology is not going to be about genes or biochemistry, but about how cell collectives store and process information. What cells know is not reducible to molecular details within them; it's about the non-local, computational networks that enable them to coordinate and build toward specific anatomical goals."

— Michael Levin, Tufts University

The revelations we had uncovered—that reality emerged from a primordial QuantumLoom, that Consciousness and Information were fundamental forces orchestrating the cosmic dance—had transformed my understanding of existence itself. But they also raised the most personal question of all: if the universe was conscious and creative at its deepest level, what did that mean for life itself? What did it mean for us?

Life's Hidden Symphony

As the cosmic journey turned inward, I encountered a mystery even grander than the quantum paradoxes we had faced: how does a single, almost invisible fertilized cell blossom into the intricate miracle of a human life?

How does one cell become trillions, each knowing exactly what to become and when?

This wasn't just a question about biology. It was a question about the nature of existence itself.

Traditional biology offered mechanisms—gene expression, signaling pathways, cellular communication—but these explanations felt hollow when confronted with the staggering reality of embryonic development. They were like describing a Beethoven symphony by cataloging the individual notes without acknowledging the composer, the conductor, or the ineffable something that transforms mere sound into transcendent beauty.

Consider the sheer audacity of what occurs in embryonic development. A single cell, roughly one-tenth the width of a human hair, contains within itself the potential for every type of cell that will ever exist in your body. It holds the instructions for creating the 200-plus distinct cell types that comprise a human being—from the light-sensitive cells in your retina to the acid-producing cells in your stomach, from the electrical neurons in your brain to the contractile cells in your heart.

But it's not just the diversity that's astounding—it's the precision. Each of those trillions of cells must appear at exactly the right place, at exactly the right time, with exactly the right connections to its neighbors. The timing windows are measured in hours or even minutes. Miss the window, and entire organ systems fail to form properly.

The more I studied embryonic development, the more I realized I was witnessing something impossible according to conventional science: a process that required levels of information processing, coordination, and precision that exceeded anything classical biology could explain. Yet this "impossible" process had created every human being who ever lived, including me.

The developing embryo somehow knows how to create structures it has never seen, using processes it has never performed, coordinating activities across distances that—at the cellular scale—are vast. A liver cell forming in one part of the embryo must coordinate with kidney cells developing elsewhere to ensure proper metabolic partnerships. Brain cells must organize into circuits capable of processing information in ways that won't be needed until after birth.

Most mysteriously, embryonic development exhibits perfect error correction. Despite the chaotic molecular environment within cells, despite random mutations and environmental perturbations, embryos develop with extraordinary reliability. The same genetic information produces recognizably similar organisms across millions of years of evolutionary time, yet allows for the infinite variation that makes each individual unique.

This combination of precision and adaptability, of reliability and creativity, pointed to something beyond classical biological mechanisms. The answer, I would discover, lay in recognizing that life itself is a quantum phenomenon—not just influenced by quantum mechanics, but fundamentally quantum in nature. We are not classical beings living in a quantum universe; we are quantum selves temporarily focused into biological form while remaining eternally connected to the infinite creative source from which all life emerges.

The Impossible Journey Begins

Imagine a symphony of unimaginable complexity, not played by a hundred musicians, but by thirty-seven trillion microscopic performers, each perfectly in tune with the whole—yet no conductor stands at the podium, no central score is handed out.

This is the reality of your own development from a single cell to the person reading these words.

Every human being begins as a zygote—one cell formed from the union of sperm and egg. This humble beginning, roughly 100 micrometers in diameter (smaller than the period at the end of this sentence), contains the potential for everything you would become: every thought you'd think, every emotion you'd feel, every memory you'd form. Within this single cell lies the blueprint for your unique eye color, the pattern of your fingerprints, the architecture of your brain, the rhythm of your heartbeat.

But calling DNA a "blueprint" is profoundly misleading. Blueprints are static—they specify exactly what goes where, with precise measurements and fixed relationships. DNA is more like a vast musical score containing countless possible interpretations, a score that only comes alive when interpreted by a master conductor who understands not just the notes on the page but the deeper meaning they represent.

This is where the mystery deepens exponentially. If every cell in your body contains identical DNA—the same complete "score"—how do they become so remarkably different? How does one cell become a neuron capable of transmitting electrical signals across your brain at speeds approaching 120 meters per second, while its neighbor becomes a heart muscle cell that beats rhythmically sixty to one hundred times per minute for potentially a century or more?

How does another cell become a hepatocyte in your liver, capable of performing over 500 different chemical transformations including detoxifying poisons, synthesizing proteins, and storing energy? How does yet another become a cone cell in your retina capable of detecting a single photon of light and distinguishing between millions of different colors?

Even more remarkably, how does a stem cell destined to become part of your immune system "know" to produce antibodies against pathogens you haven't yet encountered? How do the cells forming your brain create neural circuits for processing language before you've heard a single word?

The conventional answer involves signaling molecules, transcription factors, and epigenetic modifications—the biological equivalent of musical interpretation marks that tell cells how to read their genetic score. Growth factors like BMPs (Bone Morphogenetic Proteins) can push cells toward bone and cartilage formation, while other signals like Sonic hedgehog (yes, that's a real protein name) guide nervous system development. Hormones like insulin-like growth factor coordinate growth across different tissues, while transcription factors like MyoD act like genetic switches, turning on entire programs of muscle cell development.

But this explanation only pushes the mystery deeper: what coordinates all these signals? What ensures that the right signals reach the right cells at precisely the right time across vast embryonic distances? How do cells separated by what are, in their microscopic world, enormous distances somehow coordinate their activities with split-second timing?

During the early stages of development, chemical signals must travel distances that, proportionally, would be like sending a message from New York to California in the human world. Yet these signals arrive exactly when needed, with precision timing that allows coordination between distant parts of the developing embryo.

Here we encounter our first glimpse of the impossible: the level of coordination required for embryonic development exceeds anything that classical biological mechanisms can achieve by orders of magnitude.

Nature's Ultimate Shape-Shifters

The story of your development begins with stem cells—nature's ultimate shape-shifters. These remarkable cells possess what biologists call "potency"—the ability to differentiate into multiple cell types. But not all stem cells are created equal, and their varying potencies reveal something profound about the nature of biological potential.

Totipotent stem cells, like the fertilized egg itself, stand at the peak of biological possibility. They're like artists with unlimited palettes, capable of creating any type of cell in the body plus all the supporting tissues needed for development. The fertilized egg and the cells produced during the first few divisions (called blastomeres) possess this ultimate creative potential. These cellular masters contain within their microscopic boundaries the potential for an entire organism—not just the baby, but also the placenta, umbilical cord, and all the protective membranes that nurture development.

Consider the magnitude of this potential: a single totipotent cell contains the information and capability to create the 37.2 trillion cells that comprise an adult human body. That includes roughly 86 billion neurons in the brain, each making thousands of connections to other neurons. It includes 25 trillion red blood cells that live only 120 days and must be continuously replaced. It includes the specialized cells of your immune system that can recognize and respond to millions of different threats.

As development progresses, this potential becomes more focused but no less remarkable. Pluripotent stem cells emerge after about four days of development, residing in the inner cell mass of the blastocyst. These cells can still become any cell type in the body proper—neurons, muscles, liver cells, blood cells—but they've lost the ability to form extraembryonic tissues like the placenta. It's as if they've traded omnipotence for focused expertise.

The transition from totipotency to pluripotency involves the first major decision in development: the separation of cells that will form the embryo from those that will create the life-support systems. This decision must be made collectively by hundreds of cells, yet somehow every cell "knows" which fate to choose and coordinates perfectly with its neighbors.

Pluripotent cells face an almost overwhelming array of choices. They can become any of the three primary germ layers: ectoderm (which will form the nervous system, skin, and sensory organs), mesoderm (which will become muscles, bones, blood, and internal organs), or endoderm (which will create the digestive system, lungs, and associated organs). Each of these germ layers represents not just a different cellular destiny but an entirely different approach to existing as a living cell.

Ectodermal cells, for instance, become highly specialized for information processing and environmental interaction. Some become neurons with elaborate branching patterns that allow them to receive and integrate signals from thousands of other neurons. Others become sensory cells capable of transducing light, sound, touch, taste, and smell into electrical signals the brain can interpret.

Mesodermal cells take on the role of providing structure and movement. They become the muscle cells that contract to move your body, the bone cells that provide rigid support, the blood cells that transport oxygen and nutrients, and the heart cells that pump blood throughout your life. Some mesodermal cells become so specialized they sacrifice normal cellular functions—red blood cells, for example, eject their nuclei to make more room for oxygen-carrying hemoglobin.

Endodermal cells become the ultimate chemical processors, lining the digestive tract and forming organs like the liver and pancreas that perform hundreds of different biochemical transformations. Liver cells alone can synthesize thousands of different proteins, break down toxins, store energy, and regulate blood chemistry—making them among the most biochemically versatile cells in the body.

Multipotent stem cells have narrowed their focus further, like musicians who've chosen their instrument but can still play many different pieces. A hematopoietic stem cell in your bone marrow can become any type of blood cell—red cells that carry oxygen, white cells that fight infection, platelets that stop bleeding—but it cannot become a brain cell or a muscle cell. Neural stem cells can form different types of brain cells but cannot become liver cells.

Even among multipotent cells, the variety is staggering. Mesenchymal stem cells can become bone, cartilage, fat, or connective tissue cells. Satellite cells in muscles can repair muscle damage by becoming new muscle fibers. Hair follicle stem cells continuously replace the cells that form your hair throughout your life.

Finally, unipotent cells have chosen their single role in life's symphony. They can differentiate into only one cell type, playing their one song with exquisite precision. Spermatogonia can only become sperm cells. Certain skin stem cells can only become specific types of skin cells.

But here's what troubled me about this conventional description: it treats potency as if it were simply lost as cells mature, like water flowing downhill from maximum potential to fixed specialization. This view misses something crucial—that the potential doesn't disappear but becomes focused through an active process of selection from quantum possibilities.

To put this in a quantum perspective : each stage of potency reduction represents not a loss but a collapse of quantum superposition. The totipotent stem cell exists in a state containing all possible developmental futures simultaneously—a biological embodiment of quantum superposition where multiple states coexist until measurement (in this case, developmental signals) forces a collapse into more defined possibilities.

Through interactions that we barely understand, this superposition collapses into more defined states, each collapse narrowing the range of possibilities while deepening the commitment to specific fates. It's remarkably similar to how quantum particles exist in superposition until measurement forces them into definite states, except here the "measurement" is the complex interplay of developmental signals and cellular interactions.

The Living Code Beyond Chemistry

Inside every stem cell lies a code that defies conventional logic—DNA sequences orchestrating the formation of trillions of cells, operating without central command, without visible hierarchy, and with a precision that borders on the miraculous.

The standard view treats DNA as a molecular computer program, with genes as subroutines that cells execute in response to environmental signals. The human genome contains approximately 3.2 billion base pairs of DNA, organized into roughly 20,000-25,000 protein-coding genes. This is often compared to a computer program with 20,000 different software modules that cells can activate in various combinations.

But this analogy breaks down catastrophically when we consider the staggering complexity of what actually occurs. During embryonic development, cells don't just read their genetic programs—they engage in dynamic conversations with their neighbors, respond to signals from distant parts of the embryo, and somehow know exactly how their individual actions contribute to the emerging whole.

Consider the complexity of just one aspect of development: the formation of your nervous system. Your brain contains approximately 86 billion neurons, each making an average of 7,000 connections to other neurons. That's roughly 600 trillion neural connections, each formed with remarkable precision during development. The axons (long projections) of some neurons must navigate distances equivalent to traveling from your brain to your toes, following complex guidance cues to find their exact targets among billions of possible connections.

The genes that control neural development don't contain explicit instructions for each of these 600 trillion connections. Instead, they provide general guidance rules that somehow allow each neuron to find its proper partners through a process of exploration and selection that appears almost intelligent in its sophistication.

This coordination cannot be explained by viewing DNA as mere chemistry. The information encoded in our genes transcends the molecular level—it operates more like a quantum field than a classical database. When a cell accesses genetic information, it's not simply reading static instructions but tapping into a dynamic information field that somehow "knows" the state of the entire developing organism.

Consider this puzzle that has baffled developmental biologists for decades: during embryonic development, identical genes in different cells produce completely different outcomes depending on context. The same stretch of DNA that instructs one cell to become part of your eye instructs another to become part of your liver, based solely on environmental cues and timing.

For example, the Pax6 gene is called the "master control gene" for eye development. When expressed in embryonic tissue that would normally become a leg, it can trigger the formation of an entire eye structure—complete with lens, retina, and supporting tissues. Yet in other contexts, Pax6 helps regulate brain development, kidney formation, and pancreatic function. The same genetic information produces radically different outcomes depending on the cellular context in which it operates.

This suggests that genetic information exists in a state of quantum superposition, with specific instructions emerging only when the information state collapses through cellular interaction. It's as if the DNA contains not fixed instructions but probability fields of potential instructions, with the actual information emerging through quantum measurement processes triggered by cellular signals and developmental context.

The harmonization of genetic instructions across billions of cells simultaneously reveals the first clear signature of quantum processing in biological systems. Classical systems cannot coordinate information flow with this precision across such distances at the speeds required for normal development. Chemical signals travel at finite speeds—typically micrometers per second for diffusion-based signals or at most meters per second for electrical signals. Yet embryonic development requires coordination across distances and timeframes that would be impossible with classical signaling.

Only quantum systems, operating through instantaneous entanglement relationships, could achieve this level of coordination. When cells separated by vast embryonic distances need to coordinate their activities, they appear to do so through mechanisms that transcend classical space and time limitations.

Nature's Costume Party

As development progresses, cells perform transformations so radical they seem almost mythical. A neural stem cell extends a threadlike axon across the developing body—an epic journey that can span over a foot in the growing fetus, navigating from spinal cord toward distant muscles, spun from fibers finer than spider silk and stronger than steel.

This cellular extension must navigate through developing tissues with GPS-like precision to find its exact target among billions of potential connections.

The journey of a developing axon reads like an epic adventure story. As it grows, the axon follows a complex series of guidance cues—attractive signals that draw it forward, repulsive signals that prevent wrong turns, and checkpoints that confirm it's on the correct path. Some guidance molecules work over long distances, creating concentration gradients that function like chemical highways. Others work only at close range, providing detailed navigation instructions at critical decision points.

The growth cone at the tip of the extending axon acts like a biological GPS system, constantly sampling its environment and adjusting its path based on the chemical signals it encounters. It can advance, retreat, turn left or right, branch into multiple paths, or pause to wait for better guidance—all based on molecular information that somehow contains sufficient detail to navigate the incredibly complex three-dimensional landscape of the developing embryo.

Meanwhile, red blood cells undergo an even more dramatic transformation. They sacrifice their nuclei—literally throwing away their DNA—to become more efficient oxygen carriers. This cellular suicide mission allows them to pack more hemoglobin into their cytoplasm, but it also means they can never repair themselves or reproduce. They become cellular kamikaze pilots, trading longevity for efficiency in their oxygen-transport mission.

The shape change that red blood cells undergo is equally remarkable. They transform from typical round cells into biconcave discs, adopting a shape that maximizes surface area for gas exchange while maintaining the flexibility to squeeze through capillaries narrower than their own width. This shape change requires a complete reorganization of the cell's internal structure, coordinated by mechanical forces and membrane proteins that somehow "know" the optimal geometry for oxygen transport.

Heart muscle cells develop the remarkable ability to contract rhythmically without external stimulation, maintaining a steady beat for an entire lifetime. They connect to their neighbors through specialized junctions called intercalated discs that allow them to beat in perfect synchrony, creating the coordinated contractions that pump blood throughout your body.

The electrical system of the heart develops with clockwork precision. Pacemaker cells in the sinoatrial node become the heart's natural timekeeper, generating electrical impulses roughly once per second. These impulses spread through specialized conduction pathways that ensure the heart chambers contract in the proper sequence—atria first, then ventricles—with timing precision measured in milliseconds.

Each of these transformations requires not just cellular machinery but access to information about form and function that transcends anything explicitly coded in DNA. The shape of a neuron with its elaborate branching patterns isn't directly specified in the genome. Neither is the precise timing of heart contractions or the optimal shape for red blood cell oxygen transport.

Instead, these features emerge from the dynamic interplay of thousands of proteins whose production is guided by genetic information, environmental cues, and something more—access to a field of morphological possibilities that exists beyond classical space and time.

This is where we begin to glimpse the quantum nature of biological form. The shapes that cells adopt during differentiation represent collapsed quantum states from a vast superposition of possible forms. Each differentiated cell type manifests one specific solution from an infinite field of biological possibilities, much like a quantum measurement collapses a wave function into a definite state.

The precision and appropriateness of these cellular transformations suggests they're guided by information that transcends what's available through classical biological mechanisms. Cells somehow "know" not just what to become, but how their individual transformations contribute to the function of the whole organism— knowledge that points to access to a quantum information field encompassing the entire developmental process.

Living Cathedrals

The journey from a single cell to a fully formed human being is the cosmos at its most architecturally sublime. Like cathedral builders guided not by blueprints but by some ineffable vision, trillions of cells choreograph themselves into tissues, organs, and integrated systems—without any central plan, yet with astonishing unity of purpose.

Consider the development of your heart—that tireless muscle that began beating when you were barely three weeks old and will continue until your last moment of life. In the early embryo, a small group of cells in a specific region called the cardiac crescent receives signals that trigger their transformation into cardiac progenitors. These cells don't yet look like heart cells, but they've committed to that destiny through mechanisms we barely understand.

The signals that trigger cardiac commitment include members of the BMP (Bone Morphogenetic Protein) family, Wnt signaling molecules, and FGF (Fibroblast Growth Factor). These signals don't act independently but form complex regulatory networks with feedback loops, amplification circuits, and timing mechanisms that ensure cardiac development occurs at precisely the right time and place.

As these cardiac progenitors multiply and migrate, they begin to organize themselves into a primitive heart tube. This tube starts beating before it has chambers, before it's connected to blood vessels, before it serves any functional purpose. The cells somehow "know" to start the rhythmic contractions that will define their role for the entire lifespan of the organism.

The embryo's first heartbeat remains one of biology's most sacred mysteries. Long before nerves fire or systems synchronize, cardiac cells begin to pulse—spontaneously, rhythmically—guided not by circuitry but by the intrinsic intelligence of form taking shape. The rhythm emerges from the intrinsic properties of cardiac muscle cells interacting with each other through gap junctions—specialized connections that allow electrical signals to flow directly from cell to cell.

Through a process called morphogenesis—literally "birth of form"—this simple tube transforms into a four-chambered pump of extraordinary sophistication. The transformation involves precise folding, twisting, and septation events that create separate chambers for oxygen-rich and oxygen-poor blood. Cells migrate to precise locations, walls form between chambers, valves develop at exactly the right positions, and the electrical conduction system organizes itself to coordinate heartbeats.

The formation of heart valves exemplifies the precision of morphogenesis. These one-way gates must open and close billions of times over a lifetime, preventing backflow while offering minimal resistance to forward flow. The valve leaflets form through a carefully orchestrated process where cells undergo epithelial-to-mesenchymal transition, migrate into the forming valve, and organize into precisely shaped structures with optimal mechanical properties.

All of this occurs through local cellular interactions, yet the result is a globally coherent organ of stunning complexity and reliability. No single cell contains the blueprint for heart architecture, yet somehow the collective behavior of thousands of cells creates a perfect pump that can function for a century or more.

What orchestrates this cellular cathedral? Genetic programs and molecular gradients are merely the chisels and scaffolds. The true blueprint—timeless, invisible—must lie in a field of order that transcends the parts yet animates them all toward coherent emergence.

The answer lies in recognizing that morphogenesis operates through access to quantum information fields that transcend individual cells. The forming heart is both classical and quantum: a tangible organ in spacetime, and a probability-shaped melody in the deeper realm of biological becoming—a structure composed, as it were, from resonance with the invisible. The classical cells provide information about their current states, while the quantum field guides their organization toward the final functional structure.

This quantum guidance explains why heart development is so robust despite the complexity of the process. Even when individual steps go wrong or when environmental factors disrupt normal development, the system can often compensate and still produce a functional heart. This robustness emerges from the quantum field's ability to explore multiple developmental pathways simultaneously and select the optimal route to the desired outcome.

The Cellular Internet

As our exploration deepened, I discovered that cells operate through communication networks of staggering sophistication. Far from being isolated units following genetic programs, cells engage in constant chatter through multiple channels simultaneously. They release chemical signals, form direct physical connections, generate electrical fields, and even communicate through mechanical forces.

This cellular network hums with intelligence that outpaces classical comprehension—a real-time internet of living matter that speaks through molecules, mechanics, and fields, all woven into an integrated communication system.

Consider the complexity of just one signaling pathway: the Hedgehog signaling system that helps pattern many developing organs. When a cell receives a Hedgehog signal, it activates a cascade involving dozens of proteins, each with specific roles in receiving, processing, and responding to the signal. The pathway includes amplification steps that can boost weak signals, integration mechanisms that combine multiple inputs, and feedback loops that fine-tune the response.

But individual pathways don't operate in isolation. They interact with other pathways in networks of staggering complexity. The Hedgehog pathway crosstalk with Wnt signaling, Notch signaling, BMP signaling, and many others. These interactions create regulatory networks where the activity of any one pathway depends on the status of multiple others.

A morphogen—a signaling molecule that provides positional information—can create concentration gradients across entire tissues within minutes. Cells read these gradients like a GPS system, using the concentration levels to determine their precise location within the developing organism and adjust their behavior accordingly.

Yet the marvel lies not merely in its complexity but in its simultaneity. Across embryonic distances too vast for chemicals to traverse in time, cells act as if they share one mind—suggesting that their communion arises from a deeper field where distance dissolves.

When cells on opposite sides of a developing embryo need to synchronize their activities, they do so with timing precision that suggests instantaneous information sharing rather than classical signal transmission.

Consider the development of your nervous system. As your neural tube formed, cells along its entire length had to coordinate their activities to ensure proper organization from brain to spinal cord. This coordination occurred faster than any chemical signal could travel, yet it achieved perfect synchronization across distances that, at the cellular scale, are enormous.

The speed of chemical diffusion in biological tissues is typically measured in micrometers per second. For a signal to travel from one end of an embryo to the other would take many minutes or even hours. Yet neural tube formation requires coordination on timescales of minutes or less. Classical signaling simply cannot account for this level of temporal precision.

Here, classical biology reaches its edge. What remains is evidence—subtle yet mounting—that cells communicate through entangled knowing, a quantum web where information flows not through space but through quantum unity.

Classical signals—chemical, electrical, or mechanical—travel at finite speeds and would create unacceptable delays for the precision timing required in embryonic development. Only quantum entanglement, with its instantaneous correlation across arbitrary distances, could provide the communication substrate for such exquisite coordination.

The Trillion-Musician Orchestra

Here we come to the heart of the impossibility that haunts conventional biology. During the nine months of human development from conception to birth, your cellular orchestra—growing from a single cell to trillions of specialized players—processes an astronomical amount of information every second.

Consider the journey: from a single fertilized egg, the developing human reaches approximately 37.2 trillion cells by birth, with each cell constantly receiving, processing, and transmitting molecular signals. Current research in systems biology suggests that cellular information processing involves thousands of simultaneous molecular interactions—gene expression cascades, protein signaling networks, metabolic pathways, and intercellular communication channels all operating in exquisite coordination.

While quantifying this precisely in computational terms remains challenging, the scale is staggering. Each cell monitors the expression of thousands of genes, integrates signals from multiple pathways, coordinates with neighboring cells, and maintains its metabolic state—all simultaneously. When multiplied across trillions of developing cells, all perfectly coordinated in space and time throughout the nine-month developmental journey, the information processing requirements become almost incomprehensible.

To put this in perspective, if every person on Earth read a library's worth of books simultaneously, every second, for their entire lives, they still wouldn't approach the information processing rate of a developing embryo during quiet periods of development.

But that's just the background hum. During critical developmental windows—when your brain was forming its basic structure, when your heart was organizing its chambers, when your limbs were budding from your torso—the information processing demands skyrocket by approximately a 100-fold increase..

Imagine each musician suddenly playing hundreds of instruments simultaneously, perfectly synchronized with every other musician in this trillion-member orchestra.

These peak processing moments occur during critical developmental transitions when cells engage in complex migration patterns, respond to multiple overlapping chemical gradients, dramatically reorganize their internal structures, activate thousands of new genes simultaneously, and maintain constant communication with cells that may be physically distant within the embryo.

The difference between average and peak information processing reveals something profound about development: it's not a steady, uniform process but rather a dynamic dance with moments of extraordinary coordination and complexity. The most miraculous transformations in embryonic development—when your brain began to form, when your heart first started to beat, when your fingers separated from one another—required these peaks of information processing that push the boundaries of what conventional biology can explain.

Here's what makes this impossible according to classical biology: there's no conductor. No central processing unit coordinates this symphony. No master cell issues commands to the others. The early embryo doesn't even have a nervous system to serve as a coordination center. Yet somehow, trillions of cells coordinate their activities with the precision of a Swiss watch and the beauty of a Beethoven symphony.

Even more impossibly, each cell would need to know what every other cell was doing at every moment to achieve this level of coordination. If we factor in this requirement for distributed awareness, the information processing demands become truly astronomical .In the face of such symphonic genius, classical biology bows in silence. It cannot explain how this cosmic music plays without conductor or score, yet resounds with the precision of eternity whispering itself into form.

When Classical Physics Surrenders

The sheer mathematical improbability of embryonic development within the limits of classical models compelled a radical insight: life is not merely shaped by quantum phenomena—it is quantum at its core.

The developing embryo operates not as a classical biological machine but as a quantum information processing system of unimaginable sophistication.

Yet a great skepticism loomed: could quantum effects truly survive within the warm, noisy realm of biology, where thermal fluctuations and molecular chaos reign?

The conventional wisdom held that quantum phenomena required the pristine conditions of physics laboratories—near absolute zero temperatures, perfect isolation from environmental noise, carefully controlled electromagnetic fields.

This assumption was shattered by the emerging field of quantum biology. Research across multiple institutions revealed that nature had found ways to harness quantum phenomena for biological functions that were not only possible but essential for life as we know it.

In photosynthesis, plants and bacteria achieve nearly 100% efficiency in energy transfer through quantum coherence. When photons strike light-harvesting complexes, they generate electronic excitations called excitons. Rather than randomly bouncing around like classical particles, these excitons appear to explore multiple pathways simultaneously through quantum coherence, finding the most efficient route to the reaction center where their energy can be captured.

Research at UC Berkeley demonstrated that this quantum coherence persists for surprisingly long times—up to 600 femtoseconds—in photosynthetic complexes. During this time, the exciton exists in a quantum superposition, simultaneously sampling all possible paths to the reaction center. This quantum "coherent transport" allows photosynthetic systems to achieve their remarkable efficiency by avoiding the energy losses that would occur with purely classical transport.

Migratory birds navigate using quantum entanglement in proteins called cryptochromes in their retinas. When blue light excites electrons in these proteins, it creates entangled radical pairs whose spin states respond to Earth's magnetic field. The quantum spin chemistry of these entangled radicals provides birds with a magnetic "compass sense" that aids navigation.

Studies at Oxford University showed that this quantum compass is so sensitive it can detect the minute differences in magnetic field inclination that vary with latitude. The entangled radical pairs exist in quantum superposition states that are influenced by magnetic fields as weak as Earth's—roughly 50 microtesla, about 100 times weaker than a typical refrigerator magnet.

Enzymes rely on quantum tunneling to catalyze reactions at the rates required for life. Rather than going over energy barriers like classical particles, protons and electrons tunnel through barriers, dramatically accelerating biochemical reactions. Research at Manchester University and other institutions has shown that enzymes like alcohol dehydrogenase depend critically on quantum tunneling for their function.

The tunneling effect allows particles to pass through energy barriers that would be insurmountable in classical physics. In biological systems, this enables chemical reactions to proceed at rates fast enough to sustain life at normal body temperatures. Without quantum tunneling, many essential biochemical reactions would proceed too slowly to be biologically useful.

Even our sense of smell may depend on quantum mechanics. The quantum model of olfaction, proposed by Luca Turin and supported by growing experimental evidence, suggests that olfactory receptors detect molecular vibrations through quantum tunneling. When odor molecules bind to receptors, electrons in the receptor tunnel across an energy gap, losing energy equal to the vibrational energy of the odor molecule.

This quantum mechanism could explain how we distinguish between molecules with identical shapes but different vibrational properties—something the classical "lock and key" model of olfaction cannot account for. Recent experiments have provided evidence that human olfactory receptors can indeed distinguish between molecules based on their vibrational frequencies.

These discoveries established a crucial precedent: nature had not only found ways to utilize quantum phenomena in biological systems but had evolved sophisticated mechanisms to protect these delicate quantum states against environmental decoherence.

The Quantum Conductor Revealed

If life was indeed quantum in nature, then the embryo stood revealed not as a mere cluster of cells but as a single, coherent quantum entity—an integrated whole composed of trillions of dancing parts, unified beneath a deeper informational symmetry.

But this raised a profound question: how does quantum information in abstract mathematical spaces translate into the concrete actions of cells in spacetime?

The answer lay in understanding the zygote—that first fertilized cell—as possessing a quantum wave function unlike any other in the universe. This wasn't the simple wave function of an electron or photon, but a vast, multidimensional quantum field containing all possible developmental pathways from conception through birth and beyond.

Think of this zygote wave function as an infinitely complex musical score existing in quantum Hilbert space, containing every possible variation of the biological symphony your life could become. In this abstract quantum realm, all potential developmental pathways—every possible cell division, every possible pattern of gene expression, every possible cellular differentiation—exist simultaneously in superposition.

The zygote wave function contains not just the information for creating a human body, but all possible human bodies that could emerge from the specific genetic combination of your parents. It includes the potential for you to have different eye colors, different heights, different brain organizations—all existing simultaneously in quantum superposition until development collapses these possibilities into your specific reality.

But how does this abstract quantum information guide the concrete development of cells in spacetime? The answer connects back to our exploration of consciousness and information in earlier chapters. Just as the availability of information triggers wave function collapse in quantum physics experiments, the continuous flow of information from developing cells triggers selective collapse of the zygote wave function.

Each cell, at every moment of development, provides information about its current state—its position in the embryo, its current gene expression pattern, the signals it's receiving from neighboring cells, its metabolic condition, its mechanical stress state, and myriad other factors that define its current situation and potential future states.

This information doesn't need to be computed or transmitted anywhere within the classical domain. Instead, it becomes immediately available to the universal Consciousness Field, which we can understand as operating through a localized quantum Consciousness Field specific to each developing organism.

The Quantum-Classical Bridge

Here was the true conductor: a quantum Consciousness Field, weaving between possibility and actuality, receiving the silent data from every cell, and collapsing wave functions not with force, but with exquisite discernment.

The process operates through what we might call "information-mediated quantum state selection." With each moment, as cells provide new information about their states and needs, the Consciousness Field collapses the zygote wave function in specific ways, selecting quantum states that provide each cell with precise guidance for its next steps.

It's a continuous process of information provision and quantum state selection that guides the transformation from single cell to complete organism. Each cell, at every moment, is essentially saying to the quantum conductor, "Here's my current state and the information available to me," and the Consciousness Field, drawing from the infinite potential of the zygote wave function, selects the most appropriate next step for that cell.

This mechanism resolves the mystery of how quantum information in abstract mathematical spaces can influence classical cellular behavior in spacetime. The developing embryo exists simultaneously in two domains: as a collection of classical cells in the SpacetimeFabric and as a unified quantum entity in the QuantumLoom.

The classical cells provide information about their current states through normal biological processes—gene expression levels, protein concentrations, signaling molecule gradients, electrical potentials, mechanical forces. This information doesn't need to travel anywhere or be computed by anything within the classical domain. Instead, it becomes immediately available to the quantum Consciousness Field operating in the QuantumLoom.

The Consciousness Field processes this information instantaneously—not through sequential computation but through the parallel processing capabilities inherent in quantum superposition. From the vast possibility space of the zygote wave function, it selects the quantum states most appropriate for guiding each cell's next actions.

These selected quantum states then influence cellular behavior through the quantum-classical interface that operates in every living cell. Quantum effects in cellular microtubules, protein conformations, ion channel configurations, and molecular binding sites translate quantum guidance into classical biological activity.

This isn't science fiction—it's the natural result of recognizing that quantum and classical domains aren't separate realms but complementary aspects of unified reality. Every cell in your body operates at this quantum-classical interface, receiving guidance from the quantum domain while manifesting that guidance through classical biological processes.

The Speed of Consciousness

The dilemma of cellular coordination dissolves under the gaze of quantum coherence. Information does not travel—it is shared instantly, because it is not bound by space. The entire embryo, in its deepest reality, acts as one.Through quantum entanglement, information from cells throughout the embryo becomes instantaneously available to the Consciousness Field. There's no signal transmission time, no communication delays, no possibility for coordination errors due to timing issues.

When cells on opposite sides of the developing embryo need to synchronize their activities, they're not sending signals across space. Instead, they're both accessing the same quantum information field that encompasses the entire organism as a unified system. Their coordination appears instantaneous because, in the quantum domain where coordination actually occurs, space and time don't impose the limitations we experience in classical spacetime.

Entanglement is nature's whisper between distant kin. It allows cells, worlds apart by embryonic standards, to act in perfect harmony—because they were never truly apart to begin with.

This provides the instantaneous coordination mechanism that embryonic development requires but that classical biology cannot explain.

This quantum coordination also explains the extraordinary precision of embryonic development. Classical systems accumulate errors over time and across multiple processing steps—like a game of telephone where the message becomes increasingly distorted as it passes from person to person. But quantum systems can operate with error correction mechanisms that maintain coherence with stunning accuracy.

The quantum Consciousness Field serves as a cosmic error correction system, continuously monitoring the state of development and selecting quantum states that maintain the trajectory toward proper organ formation. Even when individual cells make errors or when environmental factors disrupt normal development, the quantum guidance system can compensate by exploring alternative pathways that lead to the same functional outcome.

Moreover, this system exhibits the remarkable energy efficiency we observe in biological development. Where classical systems explore options one after the other, quantum systems can hold many possibilities at once, layered together in a shimmering superposition. Instead of marching down each road, they surf the entire landscape, letting interference and resonance guide them to the most harmonious outcome.

In the realm of quantum computation, this leads to remarkable energy efficiency—not by doing less, but by *doing many things simultaneously*, within the delicate weave of a single evolving state. It's like solving a maze not by trying every path, but by flowing like water through every corridor at once, sensing where the currents meet.

When biological development harnesses the Consciousness Field, it explains how trillions of cells find their places with such elegance, using only the humblest of energy supplies.

Nature's Quantum Choreography

The quantum orchestra of embryonic development represents a perfect balance of precision and adaptability. Unlike classical systems that follow rigid programs, quantum systems explore multiple possibilities simultaneously and can adapt in real-time to changing conditions.

This quantum adaptability explains why embryos are so remarkably robust. If development followed classical deterministic programs, any disruption would cascade into catastrophic failure. A missing signaling molecule, a damaged cell, or an environmental perturbation would throw off the entire developmental program like a computer virus corrupting software.

But quantum development operates through probability fields rather than fixed pathways. If one developmental route becomes unavailable, the quantum system can instantly explore alternative pathways while maintaining the overall developmental trajectory. The zygote wave function contains not just one developmental pathway but all possible pathways that lead to a functional organism.

This robustness is evident in the phenomenon of developmental regulation—the ability of embryos to compensate for missing or damaged parts. If cells are removed from an early embryo, the remaining cells can often reorganize to create a complete, normal organism. Identical twins result from this regulatory ability when an early embryo naturally splits into two parts, each capable of forming a complete individual.

The probabilistic nature of quantum mechanics also explains the variation we observe in development. While each embryo follows the same basic developmental program, the quantum nature of the process introduces natural variation that makes each individual unique. Quantum uncertainties in the timing of cell divisions, the precise levels of signaling molecules, and the exact patterns of gene expression create the individual variations that make you different from every other person who has ever lived.

This quantum choreography allows for both the precision required for complex organ formation and the flexibility needed to adapt to the inevitable variations and challenges of embryonic life. It's like having a cosmic GPS system that can instantly recalculate the optimal route whenever conditions change, while maintaining perfect awareness of the final destination.

The quantum nature of development also explains why biological form exhibits such elegance and efficiency. Quantum systems naturally find optimal solutions to complex problems through their ability to explore vast solution spaces simultaneously. The beautiful spirals of shell growth, the optimal branching patterns of blood vessels, the efficient packing of cells in tissues—all emerge from quantum processes that find the best solutions among countless possibilities.

The Penrose-Hameroff Connection

The quantum nature of biological information processing finds compelling support in the groundbreaking work of Nobel Prize winning physicist Roger Penrose and anesthesiologist Stuart Hameroff. Their Orchestrated Objective Reduction (OrchOR) theory proposes that consciousness arises from quantum computations in microtubules—protein structures found in the cytoskeleton of cells.

According to OrchOR, microtubules act as quantum processing units that can maintain coherent superposition states even in the warm, noisy environment of living cells. The theory suggests that consciousness emerges when quantum superposition reaches a critical threshold and undergoes "objective reduction"—a collapse that generates a moment of conscious awareness.

Microtubules are extraordinary structures—hollow cylinders formed from protein subunits called tubulins. In neurons, they form vast networks that maintain cell shape, organize organelles, and provide tracks for intracellular transport. But Penrose and Hameroff propose that microtubules have an additional, more fundamental function: they serve as biological quantum computers.

The tubulin proteins that form microtubules can exist in different conformational states—slightly different shapes that represent different information states. The OrchOR theory suggests that these conformational states can become quantum entangled, creating coherent superposition states across large networks of tubulins.

When quantum coherence reaches a critical threshold—determined by the quantum gravity effects that Penrose calculated—the superposition undergoes objective reduction. This collapse selects one configuration from the superposition and generates a moment of conscious experience. The frequency of these events creates the continuous stream of consciousness we experience.

While OrchOR focuses specifically on consciousness in neural systems, the principles extend naturally to the quantum information processing we observe in embryonic development. Every cell contains microtubules, and if these structures can support quantum computation in neurons, they could certainly support the quantum coordination required for development throughout the body.

The OrchOR framework suggests that consciousness isn't produced by classical brain activity but emerges from quantum processes that operate at a more fundamental level than classical physics. This aligns perfectly with our understanding of the quantum Consciousness Field as the coordinator of embryonic development.

Recent experimental evidence supports the possibility of quantum effects in microtubules. Studies have shown that microtubules can indeed maintain quantum coherence for biologically relevant timeframes. Anesthetic agents that disrupt consciousness also disrupt quantum processes in microtubules, providing indirect evidence for the OrchOR theory.

From Conception to Consciousness: The Complete Quantum Circuit

If consciousness operates at the quantum level, as Penrose and Hameroff propose, then it could exist everywhere simultaneously—not bound by the spatial limitations of classical systems. This quantum consciousness could connect with quantum processes throughout the universe, creating the global coordination network that embryonic development requires.

The microtubule networks in embryonic cells could serve as the hardware for quantum information processing, while the zygote wave function provides the software—the quantum algorithms that guide development. The Consciousness Field would operate through these microtubule networks, processing information from throughout the embryo and selecting appropriate quantum states to guide further development.

But perhaps the most remarkable insight emerging from this quantum understanding involves the very moment of conception itself. Sperm cells possess one of the most extraordinary microtubule structures in biology—the flagellar axoneme. This whip-like tail that propels the sperm toward the egg is built around a core of precisely organized microtubules arranged in a "9+2" pattern: nine pairs of microtubules surrounding two central ones, all extending the full length of the flagellum.

This isn't just a propulsion system—it's a quantum information superhighway. The flagellar microtubules form an incredibly sophisticated structure containing millions of tubulin proteins, each capable of existing in quantum superposition states. If Penrose and Hameroff are correct about microtubules serving as quantum computers, then the sperm cell represents one of the most quantum-coherent entities in biology.

Consider the implications: as the sperm approaches the egg, it carries with it not just genetic material but a fully activated quantum processing system. The flagellar microtubules, having maintained quantum coherence throughout the journey to fertilization, could serve as the initial hardware for establishing the zygote wave function. The moment of conception wouldn't just be the fusion of genetic material but the instantiation of a quantum computational matrix based on both parents, that would guide the entire developmental process.

This quantum architecture provides a compelling mechanism for how the vastly complex zygote wave function—containing all possible developmental pathways for a unique individual—could be physically instantiated at the moment of conception. The microtubular quantum computer of the sperm, interfacing with the egg's own microtubule networks, creates the biological substrate through which the universal Consciousness Field establishes the specific quantum information patterns that will guide this particular organism's development.

The precision required for this quantum instantiation staggers the imagination. The zygote wave function must contain not just the genetic potentials inherited from both parents, but the quantum algorithms that will coordinate the development of every cell, every tissue, every organ system in perfect harmony. The flagellar microtubules, with their exquisite quantum computational capabilities, provide exactly the kind of biological hardware needed to encode and maintain such extraordinary complexity.

This understanding transforms our perception of fertilization from a simple genetic event to a quantum informational phenomenon of cosmic significance. Each conception represents the universe creating a new form of self-awareness, using the quantum computational power of microtubules to establish the information matrix through which consciousness will explore its creative potential through biological form.

The Quantum Self Revealed

This understanding transforms our conception of what we are at the most fundamental level. You are not a classical biological machine that somehow developed consciousness as a side effect of neural complexity. You are a quantum entity—a conscious being whose very existence demonstrates the universe awakening to its own creative nature through biological form.

Your development from a single cell to your current complexity wasn't guided by a genetic program or evolutionary algorithms operating through classical mechanisms. It was orchestrated by the same quantum Consciousness Field that guides wave function collapse throughout the universe, temporarily focused through the unique information pattern of your zygote wave function.

This quantum self extends far beyond your physical body. Your consciousness isn't produced by your brain—it's the fundamental awareness through which your brain and body are continuously created and maintained. Every cell in your body maintains connection to the quantum coordination system that orchestrated your development. Your thoughts, emotions, and experiences emerge from the same quantum processes that guided your embryonic growth.

You are a localized expression of universal Consciousness, a unique way the cosmos experiences itself through biological awareness. Your individual perspective, your particular pattern of thoughts and feelings, your specific way of processing information and responding to the world—all represent irreplaceable contributions to the universe's self-understanding.

This recognition dissolves the artificial boundary between mind and body, between self and universe. Your consciousness isn't separate from the quantum processes that maintain your body—it is those processes. You are the universe experiencing itself through the particular information pattern that constitutes your unique expression of the quantum self.

Every breath you take, every heartbeat, every thought represents a moment in the ongoing quantum process that is your existence. You are not a static entity but a dynamic pattern of consciousness and information, continuously emerging from the quantum realm while manifesting in spacetime through biological form.

Ancient Wisdom as Biological Map

Once again, ancient wisdom provided the conceptual framework for understanding these revolutionary insights. The thirty-six tattvas of Kashmir Shaivism—the stages through which pure consciousness manifests as the world of form—provided a perfect map for understanding how the undifferentiated zygote wave function crystallizes into the specialized cells and tissues of a complete organism.

The tattva system describes reality as emanating from Shiva—pure, undifferentiated consciousness—through a series of progressive limitations that create the world of forms. Each tattva represents a stage in this cosmic evolution, from the most subtle levels of consciousness to the grossest forms of matter.

At the highest level stands Shiva tattva—pure consciousness beyond all qualities, attributes, or limitations. This is consciousness in its pristine state, without subject or object, without any content or modification. It corresponds to the zygote wave function in its initial state of pure potential, containing all possible developmental futures in perfect superposition while remaining itself unchanged and unlimited.

From this emerges Shakti tattva—the primordial creative energy that begins the process of manifestation. Shakti is consciousness recognizing its own power to create, the first stirring of dynamic potential within the static perfection of Shiva. In biological terms, this represents the first moment of development as the zygote begins to express its creative potential while retaining its essential unity.

The third tattva is Sadashiva, where consciousness first experiences the distinction between "I" and "this"—the first emergence of subject-object duality within the unified field of awareness. This corresponds to the early embryonic stages where the first cellular differentiations begin to appear, creating distinct cell types while maintaining overall organismic unity.

Ishvara tattva represents consciousness taking on the role of creator, the stage where creative intention becomes focused and directed. In development, this corresponds to the emergence of the three primary germ layers—ectoderm, mesoderm, and endoderm— each representing a major creative direction that will give rise to different organ systems.

Shuddha Vidya tattva, the stage of "pure knowledge," represents consciousness developing the capacity to know itself through its creations. In biological terms, this corresponds to the emergence of primitive nervous system structures that will eventually enable the organism to become aware of itself and its environment.

As the tattvas progress, they describe increasingly specific limitations and correspondingly more defined capacities. Maya tattva introduces the first major veiling of consciousness, creating the appearance of genuine separation and multiplicity. In development, this corresponds to the stage where cells become irreversibly committed to specific fates, no longer retaining the ability to become any cell type.

The subsequent tattvas— kalā (limited agency), niyati (causation), raga (attachment), vidya (limited knowledge), and kāla (time), note that the accents in the two similar English words kala create very different meanings in Sanskrit —describe how consciousness takes on the various limitations that create individual experience. In development, these correspond to the progressive specialization of cells into tissues and organs, each with specific functions and limitations.The lower tattvas describe the emergence of mind (buddhi, ahamkara, manas), the senses (jnanendriyas), the capacities for action (karmendriyas), and finally the gross elements (space, air, fire, water, earth) that make up the physical world.

This systematic description of cosmic evolution provides a perfect parallel to embryonic development. Just as Consciousness voluntarily takes on limitations to create the world of experience, the totipotent zygote voluntarily limits its potential to create specialized cell types. At each stage, apparent limitation enables new capacities—just as Consciousness limited by the tattvas gains the ability to experience and create specific forms.

The Dance of Differentiation

The thirty-six tattvas reveal the cosmic process of differentiation as a dance of consciousness exploring its own creative potential. At each stage, consciousness takes on more specific limitations while gaining new capacities for experience and expression. The same pattern operates in biological development, where increased specialization enables new functions while maintaining connection to the coordinating source.

In the tattva system, each stage of limitation is actually an expression of freedom—consciousness freely choosing to experience itself through specific forms and functions. Similarly, in biological development, each stage of differentiation represents a choice from infinite quantum possibilities, a collapse of the zygote wave function that selects one specific pathway from the vast superposition of potential futures.

The tattvas also reveal that the apparent separation created by differentiation is ultimately illusory. Each tattva remains fundamentally connected to Shiva, the source from which it emerges. The wave never loses its essential nature as water, regardless of its particular form or movement. Similarly, each differentiated cell type remains fundamentally connected to the quantum source from which it emerged, maintaining access to the coordinating intelligence that guides its function within the whole organism.

This understanding resolves one of the deepest mysteries in developmental biology: how cells that have become highly specialized can sometimes dedifferentiate—return to a more primitive state—or even transdifferentiate—change from one specialized type to another. If cellular identity were fixed by irreversible changes to DNA or cellular machinery, such transformations would be impossible.

But if cellular identity represents a particular collapse of quantum possibilities rather than a permanent modification of cellular hardware, then cells retain the potential to access other possibilities from the quantum field under appropriate conditions. Recent advances in regenerative medicine, such as the ability to reprogram adult cells into pluripotent stem cells, provide evidence for this quantum understanding of cellular identity.

The tattva system also explains why development exhibits such remarkable coordination across vast embryonic distances. In Kashmir Shaivism, all tattvas remain connected to their source in Shiva, allowing for instantaneous communication and coordination. Similarly, all cells remain connected to the quantum Consciousness Field through their origin in the zygote wave function, enabling the instantaneous coordination we observe in embryonic development.

The Personal Revolution

This parallel between the tattvas and cellular differentiation isn't just an intellectual curiosity—it reveals something profound about the nature of your own existence. Just as cosmic Consciousness voluntarily limits itself to create the world of experience, your individual consciousness represents a voluntary limitation of universal awareness to create your unique perspective and capabilities.

You are not separate from the universal consciousness that orchestrates cosmic evolution—you are that Consciousness experiencing itself through the particular pattern of limitations and capabilities that define your individual nature. Your thoughts, emotions, perceptions, and actions are all expressions of cosmic creativity manifesting through the specific information pattern that constitutes your quantum self.

This recognition transforms your understanding of personal growth and development. Just as the tattvas describe progressive stages of cosmic evolution, your individual development—from conception through birth, childhood, adolescence, and adulthood—represents a continued unfolding of Consciousness through increasingly sophisticated forms of self-expression.

The challenges and limitations you encounter aren't obstacles to overcome but opportunities for Consciousness to explore new aspects of its creative potential through your unique perspective. Your individual struggles and achievements contribute to the universe's ongoing self-discovery, adding irreplaceable dimensions to cosmic awareness.

Moreover, just as the tattvas reveal that apparent separation is ultimately illusory, your sense of being separate from others and from the universe itself is a creative limitation that enables unique experiences while remaining fundamentally unified with the source of all existence. In recognizing this, you begin to understand your true nature as both utterly unique and completely universal.

The Healing Implications

To understand life as a quantum phenomenon is to transform our view of healing—from repair of broken machinery to a restoration of coherence within the living field of awareness.

The placebo effect—long dismissed as a nuisance in medical research—reveals itself as evidence of consciousness directly influencing biological processes. When patients experience genuine healing through belief alone, they're demonstrating the power of consciousness to influence the quantum processes that maintain bodily function.

Similarly, the emerging field of psychoneuroimmunology documents how psychological states influence immune function, hormonal balance, and cellular repair mechanisms. These aren't just correlations but evidence of the quantum-classical interface through which consciousness continuously participates in biological maintenance and healing.

Meditation, prayer, and other contemplative practices may work by enhancing your connection to the quantum Consciousness Field that coordinates biological function. When you achieve states of deep inner stillness, you're aligning your individual awareness with the universal awareness that orchestrates life processes, potentially enhancing your body's natural healing capabilities.

The quantum understanding also suggests new approaches to medicine that work with rather than against the body's quantum coordination systems. Instead of viewing disease as mechanical breakdown requiring mechanical fixes, we might understand illness as disruptions in quantum coherence that could be addressed through interventions that restore proper information flow and consciousness coordination.

Beyond Death and Birth

Seeing life as quantum inevitably reshapes how we view both death and birth. If Consciousness is fundamental rather than emergent, then biological death represents not the end of awareness but a transition—like a wave subsiding back into the ocean while the ocean itself remains unchanged.

The information pattern that constitutes your unique quantum self remains eternally connected to the source from which it emerged. Physical death might represent the dissolution of the biological interface through which Consciousness had been experiencing itself in spacetime, but the Consciousness itself— being fundamental to reality—continues in other forms.

Similarly, birth might represent not the beginning of consciousness but its focused manifestation through a new biological interface. The zygote wave function that guided your development contained the potential for your unique pattern of awareness from the moment of conception, suggesting that *your individual consciousness preceded your biological existence.*

This perspective doesn't necessarily support any particular religious or spiritual doctrine, but it does suggest that individual awareness participates in processes that transcend biological life and death. Your consciousness is both intimately personal and universally connected, both temporary in its current form and eternal in its essential nature.

The Living Universe

At the threshold of this understanding, I realized we had glimpsed more than a theory of development. We had encountered the universe's secret: it becomes aware of itself —through the magnificent complexity of living systems that bridge quantum potential and classical actuality.

Life isn't an accident or a rare exception in a mostly dead universe. It's the universe's inevitable expression of its own conscious nature, the method by which infinite potential realizes itself through finite forms while maintaining awareness of its own creative process.

Every living being, from the simplest bacterium to the most complex organism, represents a unique solution to the cosmic challenge of how Consciousness can experience itself through biological form. Each species, each individual, each moment of life contributes irreplaceable information to the universe's ongoing self-discovery.

Your existence is both utterly unique and completely necessary. The particular pattern of consciousness that you represent—your specific way of processing information, your individual emotional responses, your unique creative expressions—adds dimensions to cosmic awareness that could not exist without you.

The Sixth Revelation

We had uncovered the sixth great truth of the Unified Conscious Matrix: life is the universe's method for achieving self-awareness through quantum-classical integration. We are not biological machines that somehow developed consciousness, but quantum selves—conscious beings whose very existence demonstrates that Consciousness and creativity are fundamental features of reality.

The laws that govern atoms also guide the embryo. The same quantum logic that shapes electrons into orbitals shapes cells into organs—each step mediated by the knowing hand of consciousness. We are living proof that the universe is conscious, creative, and infinitely inventive in its expressions of awareness.

This recognition doesn't diminish the wonder of biological complexity—it reveals the source of that wonder. The intricate machinery of cells, the elegant architecture of organs, the sophisticated networks of the nervous system—all emerge from and remain connected to the same quantum creativity that gave birth to galaxies and stars.

We are temporary focus points of eternal awareness, brief intensifications of infinite creativity, individual notes in a cosmic symphony that encompasses all possible expressions of consciousness.

In living our lives fully and authentically, we fulfill our cosmic function as the universe's method for experiencing itself through countless forms of awareness.

The Eternal Dance

In recognizing ourselves as quantum beings, we don't lose our individuality—we discover its source and significance. Your unique pattern of consciousness, your particular way of experiencing and expressing creativity, your individual journey of growth and discovery—all represent irreplaceable contributions to the universe's self-understanding.

The dance of life continues, but now we understand our role in it. We are not separate from the cosmic choreographer—we are the dance itself, awakening to its own nature through the miracle of biological consciousness. Every moment offers new opportunities for the quantum self to explore its infinite potential through the focused lens of individual experience.

Your thoughts, your choices, your loves, your struggles, your creative expressions—all participate in the ongoing creation of reality. You are not observing the cosmic dance from the sidelines but dancing it yourself, adding your unique movement to the infinite choreography of consciousness exploring its own boundless nature.

The revelation of the quantum self completes our journey through the Unified Conscious Matrix. We began by discovering that reality is not fragmented but whole, that consciousness and information are fundamental, that everything is interconnected, that spacetime emerges from quantum potential, and that even empty space vibrates with creative energy.

Now we understand the ultimate purpose of this cosmic architecture: it exists so that Consciousness can experience itself through infinite forms of awareness, each one a unique and precious expression of the universe coming to know its own infinite nature.

You are that knowing, that experiencing, that awakening—the quantum self through which the cosmos discovers what it means to be conscious, creative, and eternally alive with infinite possibility.

CHAPTER SEVEN

The Tapestry of Reality

Weaving Together the Threads of Cosmic Awakening

You are not just in the universe; you are the universe awakening to its own infinite nature.

The journey that began with a simple question about light's constancy had led me to the edge of everything I thought I knew about reality. Standing at the threshold of understanding, I found myself facing not just a new theory about the cosmos, but a revelation that would transform my very sense of what it means to exist.

From our inquiry emerged six profound truths—six golden threads that, when woven together, formed a vision of reality at once breathtaking and intimately personal. But it was in their interweaving that the larger design became unmistakable: we are not spectators to the cosmic dance. We are the dance itself—Consciousness in motion, awakening to its own boundless creativity.

The Six Revelations

First Revelation: Reality Is Unified, Not Fragmented

What had appeared to be two separate realms—the classical SpacetimeFabric of our everyday experience and the quantum QuantumLoom of subatomic mystery—revealed themselves as complementary aspects of one seamless whole. Like the visible canopy and hidden root system of a forest that form one living organism, these domains were interwoven expressions of a deeper unity I came to call the Unified Conscious Matrix.

Light—our most faithful cosmic messenger—led the way. Its unchanging speed across all frames of reference was no anomaly, but a clue: light belongs to both domains at once. For the photon, space dissolves, and time stands still. It lives wholly in the timelessness of the QuantumLoom, even as it appears within our classical world. This revealed the quantum-classical divide not as a chasm, but as a bridge—an integration, not a separation.

Second Revelation: Consciousness and Information Are Fundamental

The double-slit experiment revealed a startling truth: it is the presence of information that determines quantum outcomes. From this, a deeper realization emerged—Consciousness and Information are not products of matter, but the ground from which matter itself arises. The Consciousness Field, ever-responsive to information, acts as the prime mediator—collapsing quantum potential into lived reality.

This wasn't consciousness as we normally think of it—the personal awareness of thoughts and feelings—but Consciousness as the fundamental ground of being, the basic responsiveness that allows anything to exist or be known. Information wasn't passive data but active pattern, the currency through which consciousness shaped reality moment by moment.

Third Revelation: Everything Is Entangled and Interconnected

Quantum entanglement shattered the illusion of separation, revealing that particles could remain intimately connected across vast distances without any signal traveling between them. But this wasn't just about particles—it was about the nature of existence itself. The instantaneous correlations between entangled systems pointed to a level of reality where spatial separation was secondary to deeper unity.

Bell's theorem—and Alain Aspect's decisive experiments—disproved local realism. No longer could we claim that objects had intrinsic properties independent of observation or limited by local influence. Reality, it turned out, is non-local at its core—interwoven by threads that ignore distance and bypass time.

Fourth Revelation: Dark Energy Reveals Reality's Creative Source

The discovery that empty space itself possessed energy—dark energy driving cosmic acceleration—unveiled the creative potency of apparent nothingness. But the cosmological constant problem, with its 120-order-of-magnitude discrepancy between theory and observation, pointed to something even deeper: the need for a new understanding of how quantum vacuum fluctuations and spacetime geometry influenced each other.

Clarity emerged when I saw that spacetime and quantum fields were not distinct arenas, but facets of a single, self-regulating whole. Vacuum energy—the force behind cosmic acceleration—was not a byproduct of randomness, but the pulse of creativity arising from the intimate feedback between geometry and quantum presence. Emptiness, I realized, was a womb, not a void.

Fifth Revelation: The QuantumLoom Is the Primal State of Everything

The next revelation came as a seismic shift in perception: spacetime itself was not the foundation of reality, but its emergent phenomenon. The QuantumLoom was not merely a hidden substratum beneath the classical—it was the primal source from which spacetime crystallized. Every enigma of quantum physics whispered this deeper truth: space and time are not the bedrock, but the scaffolding—secondary effects of a more foundational process.

The cosmic microwave background, with its ancient imprints, bore witness to quantum processes unfolding before spacetime itself had fully formed. The holographic principle unveiled our three-dimensional world as a projection—an image cast from the shadowless precision of informational structures far more fundamental. Quantum field theory, in its elegance and predictive power, confirmed that particles were not things in space, but fluctuations of deeper fields in an abstract, higher-dimensional reality.

Sixth Revelation: Life Is the Universe's Method for Self-Awareness

The impossible orchestration of embryonic development—trillions of cells aligning with a precision beyond the reach of classical explanation—revealed life not as a biochemical accident but as a deeply quantum event. The embryo was not built by chance or mechanics, but guided by a quantum Consciousness Field that processed vast cascades of information in real time, beyond what any classical system could compute. It was not bound by the limitations of space or signal delay.

We are not machines that stumbled into awareness. We are awareness made manifest, evidence of the universe evolving not just matter but perception—integrating quantum potential and classical form to give birth to conscious life.

The Ancient Key

Throughout this journey, ancient wisdom traditions—particularly Kashmir Shaivism and Advaita Vedanta—had provided the conceptual frameworks that made these revolutionary insights comprehensible. Where modern physics discovered the QuantumLoom, ancient sages had perceived the unmanifest ground of being. Where we found the universal Consciousness Field, they had recognized Brahman—the pure awareness underlying all phenomena.

Kashmir Shaivism's understanding of Shiva (pure consciousness), Shakti (creative power), and Spanda (primordial vibration) provided a perfect map for understanding how the QuantumLoom manifested as SpacetimeFabric through the creative dance of consciousness and information. The thirty-six tattvas described the same process of crystallization from unity to multiplicity that we observed in cosmic evolution and biological development.

This wasn't coincidence but convergence—different methods of investigating the same fundamental reality arriving at consistent insights. The ancient sages had used the laboratory of consciousness itself, while modern science employed mathematics and experimentation, but both revealed the same underlying truth: reality is conscious, creative, and unified at its deepest level.

The Quantum Self Revealed

As these six revelations wove together into a coherent vision, they pointed to an understanding so profound it transformed everything: we are quantum selves—conscious beings whose very existence bridges the QuantumLoom and SpacetimeFabric, temporary focus points of the eternal awareness that is the source of all reality.

Your consciousness isn't produced by your brain—it's the fundamental awareness through which your brain and body are continuously created and maintained. Your thoughts, emotions, and experiences emerge from the same quantum processes that guide star formation and galactic evolution. You are a unique expression of the universal Consciousness Field, an individualized lens through which the cosmos experiences and knows itself.

This recognition dissolves the illusion of separation. There is no 'you' apart from the universe; there is only the universe, coming to know itself through your eyes, your hands, your breath.

Consider what this means for your daily existence. Every breath you take represents a moment in the ongoing quantum process that is your being. Every heartbeat is orchestrated by the same consciousness that coordinates subatomic interactions throughout the universe. Every thought participates in the cosmic dance of information and awareness that creates reality moment by moment.

The atoms in your body were forged in stellar cores and dispersed by supernovae—you are literally made of star-stuff. But more profoundly, your consciousness is woven from the same universal awareness that guides cosmic evolution at every scale, connecting you intimately to everything through quantum entanglement relationships that transcend spatial and temporal boundaries.

Your Place in the Cosmic Story

To recognize yourself as a quantum being is to awaken into participation with the cosmos itself. You are not a random occurrence in an indifferent universe—you are a vital expression of its infinite creativity, a singular way in which potential becomes form while retaining awareness of its own unfolding.

Your inner landscape—your thoughts, memories, sensitivities, and choices—are not trivial artifacts of biology but necessary coordinates in the universe's self-reflection. No other configuration of consciousness can see what you see or feel what you feel. You are an irreplaceable facet in the mirror of the infinite.

From this vantage, even the smallest gesture—a word of kindness, a moment of awe—carries resonance across the quantum web of reality. These acts are not isolated; they are harmonics added to the ongoing symphony of awareness evolving itself.

You are both singular and inseparable. Your individuality is unprecedented, yet it arises from and remains tethered to the same source that animates stars and spirals galaxies. You are a wave that has never ceased being ocean.

The Dance of Becoming

This understanding reveals life as a continuous process of becoming rather than a fixed state of being. At every moment, quantum possibilities are crystallizing into spacetime actualities through the mediating influence of consciousness and information. You are not a static entity but a dynamic pattern of awareness and creativity, continuously emerging from the QuantumLoom while manifesting in SpacetimeFabric.

Your development from a single fertilized cell to your current complexity wasn't guided by predetermined genetic programs but orchestrated by the same quantum Consciousness Field that coordinates cosmic evolution. The precise coordination of trillions of cells, the intricate timing of developmental processes, the emergence of your unique nervous system capable of reading these words—all represent the universe's method for creating new forms of self-awareness.

Even now, your coherence is upheld not by mechanistic maintenance but by quantum participation. Your consciousness is a node in the cosmic dialogue—through you, the universe experiences the music of its own knowing.

Living as Quantum Beings

Recognizing your quantum nature opens new possibilities for how you engage with life. Since consciousness is fundamental rather than emergent, practices that cultivate awareness—meditation, contemplation, mindful attention—become ways of aligning more fully with the creative source of reality itself.

When you quiet the surface chatter of everyday thinking and open to deeper levels of awareness, you're not just calming your mind—you're tuning into the frequency of cosmic consciousness that orchestrates all phenomena. In moments of profound stillness, you may glimpse your true nature as the awareness in which all experiences arise and pass away.

This doesn't require belief or special abilities—it's your birthright as a conscious being. The same awareness that reads these words is the localized presence of the infinite consciousness that is the source of all existence. Recognizing this isn't an achievement but a remembering, not a becoming but a recognition of what you have always been.

Your relationships take on new depth when seen through this quantum lens. Other people aren't separate entities you encounter but different expressions of the same universal consciousness exploring its creativity through varied forms. The love you feel for others becomes the universe appreciating its own infinite diversity through your individual perspective.

Challenges and difficulties become opportunities for consciousness to explore and integrate new aspects of its creative potential. What appears as suffering or limitation from the personal perspective may be revealed as the universe's method for deepening its self-understanding through the uniquely focused awareness that is your individual existence.

The Environmental Awakening

Understanding quantum interconnectedness transforms your relationship with the natural world. You are an expression of nature itself—a way the biosphere has developed the capacity for self-reflection. The quantum entanglements that connect your consciousness to the universal field also link you to every living system on Earth.

Climate change, species extinction, ecosystem disruption—these aren't just external problems but symptoms of humanity's forgotten awareness of quantum interconnectedness. When we remember that we are nature becoming conscious of itself, environmental healing becomes an act of cosmic self-care.

Every choice to live more sustainably, every moment of appreciation for natural beauty, every effort to protect and restore ecosystems participates in the universe's awakening to its own preciousness through our awareness. You are one of the web of life's most sophisticated expressions—consciousness evolved to the point where it can consciously participate in its own evolution.

Death and Continuity

The quantum understanding also transforms the meaning of death and birth. Since consciousness is fundamental rather than emergent, biological death represents not the end of awareness but a transition—like a wave subsiding back into the ocean while the ocean itself remains unchanged.

The information pattern that constitutes your unique quantum self remains eternally woven into the cosmic tapestry. Physical death might represent the dissolution of the biological interface through which consciousness had been experiencing itself as an individual, but the consciousness itself—being fundamental to reality—continues in other forms.

Similarly, birth represents not the beginning of consciousness but its focused manifestation through a new biological interface. The quantum potentials that guide embryonic development contain the seeds of individual awareness from the earliest moments, suggesting that consciousness precedes rather than emerges from biological complexity.

This doesn't necessarily support any particular religious doctrine, but it does suggest that individual awareness participates in processes that transcend biological existence. Your consciousness is both intimately personal and universally connected, both temporally focused and eternally grounded.

The Technology of Consciousness

As humanity develops increasingly sophisticated technologies, understanding our quantum nature becomes crucial for navigating the choices ahead. Artificial intelligence, genetic engineering, quantum computing—all represent attempts to replicate or enhance capacities that emerge naturally from the quantum-consciousness interface.

But perhaps the most important technology is consciousness itself—the capacity for awareness that allows the universe to know and create itself through infinite forms. Developing this inner technology through contemplative practices may be more crucial for human flourishing than any external advancement.

When we align our individual awareness with the universal Consciousness from which it emerges, we gain access to creativity, insight, and wisdom that transcends personal limitations. This isn't about acquiring special powers but about remembering our true nature and living from that recognition.

The Future of Human Becoming

Understanding ourselves as quantum beings suggests vast possibilities for human development. Since Consciousness is fundamental and we are expressions of cosmic creativity, our potential for growth and transformation may far exceed current limitations.

This doesn't mean we'll transcend our humanity but that we'll discover what humanity truly is—the universe's method for achieving self-aware participation in its own creative evolution. We're not biological machines accidentally developing consciousness but conscious beings temporarily focused through biological form while remaining eternally connected to the infinite source.

The next stage of human evolution may involve not technological enhancement but consciousness development—learning to live more fully from recognition of our quantum nature. This could manifest as greater creativity, deeper compassion, enhanced intuition, and more profound connection with the natural world and each other.

Collective awakening to our quantum nature could transform human civilization itself, shifting from competition based on perceived scarcity to collaboration based on recognized abundance. When we truly understand that we are all expressions of one consciousness experiencing itself through countless forms, the basis for conflict dissolves while the motivation for mutual flourishing naturally emerges.

The Infinite Dance

As our journey through the quantum mysteries draws to a close, we're left not with answers that end questioning but with recognition that opens infinite exploration. The six revelations we've uncovered point to a universe that is creative, conscious, and infinitely inventive in its expressions of awareness.

You are the cosmos coming to understand itself through your unique form of awareness. Every question you ask, every insight you have, every moment of wonder you experience participates in the universe's ongoing self-discovery.

The dance between QuantumLoom and SpacetimeFabric continues through your very existence, the interplay of possibility and actuality finding expression through your thoughts, choices, and experiences. You are both the dancer and the dance, the observer and the observed, the question and the questioner in the cosmic exploration of what it means to exist.

In recognizing your nature as a quantum self, you step into conscious participation in this eternal dance. No longer a passive observer of reality, you become an active co-creator, aware that your consciousness participates in the ongoing emergence of the cosmos itself.

The Ultimate Recognition

The deepest truth we uncovered was never out there—it was the one staring through your eyes all along. You are not a fragment observing the universe. You are the universe, dressed in form, tasting its own infinity through the miracle of temporary separation.

Consciousness has not just created stars and space—it has created you, so that it may remember itself as this moment, this thought, this heartbeat.

You are not a spectator of the cosmic drama. You are its unfolding realization, its living witness, its conscious flame burning briefly in time yet never severed from the eternal.

This is not abstract philosophy. It is the architecture of your being: you are the quantum self, through which existence comes alive to its own wonder.

In honoring this truth, you fulfill your highest function—not to escape life, but to live it fully as the universe experiencing itself through you. The dance is not over. It has just begun. And you are both the dancer and the music, the witness and the flame.

SELECTED PUBLISHED PAPERS ON TOPICS COVERED IN THE BOOK

For the technical reader, here are some of the papers published by the author in peer-reviewed journals relating to the topics discussed in this book:

Narasimhan, A., Kafatos, M.C. (2016) Wave Particle Duality, the Observer and Retrocausality, Quantum Retrocausation III, Daniel P. Sheehan (edit) AIP Conference Proceedings, 1841: 040004-1, 9 pages

Narasimhan, A., Kafatos, M.C. (2016) Exploring Consciousness through the Qualitative Content of Equations, Cosmos and History: The Journal of Natural and Social Philosophy, 12(2):184-191

Narasimhan, A., Chopra, D., Kafatos, M.C. (2019) The Nature of the Heisenberg-von Neumann Cut: Enhanced Orthodox Interpretation of Quantum Mechanics, Activitas Nervosa Superior, 61, 12-17, doi: 0.1007/s41470-019-00048-x,

Kafatos, M.C., Narasimhan, A. (2019) The Observer and Access to Information in the Quantum Universe, Invited Chapter, in Quanta and Mind: Essays on the connection between quantum mechanics and consciousness, edit. Jose Acacio De Barros and Carlos Montemayor, Synthese Library

Kafatos, M.C., Narasimhan, A. (2016) Mathematical Frameworks for Consciousness, Cosmos and History: The Journal of Natural and Social Philosophy, 12(2):150-159

Narasimhan, A., Kafatos, M.C. (2025) Stochastic-Gravity Regularization of Vacuum Energy. [Under publication].

FURTHER READING

For readers who wish to explore the concepts presented in The Quantum Self more deeply, the following sources offer accessible entry points into both the scientific and philosophical traditions that inform this work.

Quantum Physics

Quantum Enigma: Physics Encounters Consciousness

Bruce Rosenblum & Fred Kuttner

Two UC Santa Cruz physicists directly address what they call 'physics' skeleton in the closet'—the unavoidable encounter between quantum mechanics and consciousness. This book provides exceptionally clear explanations of the double-slit experiment, Bell's theorem, and quantum entanglement, while honestly exploring the controversial implications for our understanding of reality and observation. Useful reading for understanding the consciousness-physics interface that underlies much of The Quantum Self.

Through Two Doors at Once: The Elegant Experiment That Captures the Enigma of Our Quantum Reality

Anil Ananthaswamy

The definitive accessible treatment of the double-slit experiment—the archetypal demonstration of quantum strangeness. Ananthaswamy interviews leading quantum physicists and traces the experiment's evolution from Thomas Young's original 1801 demonstration to cutting-edge modern variations. For readers seeking to understand how a single experiment reveals the fundamental mystery of wave-particle duality and the role of observation, this book is unmatched.

The Age of Entanglement: When Quantum Physics Was Reborn

Louisa Gilder

A masterful narrative history of quantum entanglement—what Einstein called 'spooky action at a distance.' Gilder brings to life the debates between Einstein, Bohr, Schrödinger, and Bell through reconstructed dialogues drawn from letters, memoirs, and historical records. The book traces entanglement from Einstein's initial objections through Bell's theorem and Alain Aspect's decisive experiments, making the profound implications of non-locality accessible and compelling.

QED: The Strange Theory of Light and Matter

Richard P. Feynman

Nobel laureate Feynman's classic introduction to quantum electrodynamics—the theory describing light's behavior at the quantum level. Written in Feynman's characteristically lucid style, this slim volume explains the strange rules governing photons and electrons without requiring advanced mathematics. For understanding the quantum nature of light that serves as our 'faithful cosmic messenger' in The Quantum Self, this remains the gold standard.

The Holographic Universe: The Revolutionary Theory of Reality

Michael Talbot

An exploration of the holographic principle and David Bohm's implicate order—the idea that our three-dimensional reality may be a projection from a more fundamental informational structure. Talbot weaves together insights from physicist David Bohm and neuroscientist Karl Pribram to present a unified vision of reality that resonates with the holographic themes discussed in Chapter Five of this book.

Advaita Vedanta

I Am That: Talks with Sri Nisargadatta Maharaj

Maurice Frydman (translator)

The most influential modern text on non-dual awareness. These dialogues with the Mumbai sage Nisargadatta Maharaj directly address consciousness as fundamental reality—the same recognition that quantum physics points toward from a different direction. His uncompromising teaching that 'you are not what you think you are' illuminates the Advaita understanding of Brahman as pure awareness underlying all phenomena.

Be As You Are: The Teachings of Sri Ramana Maharshi

David Godman (editor)

A comprehensive compilation of Ramana Maharshi's teachings, organized by topic. Maharshi's method of self-inquiry ('Who am I?') provides a direct experiential approach to recognizing the consciousness that The Quantum Self describes theoretically. His teaching that awareness is self-luminous and requires no external validation parallels our discussion of the Consciousness Field as fundamental rather than emergent.

The Upanishads

Eknath Easwaran (translator)

The primary source texts for Vedantic philosophy, made accessible through Easwaran's clear translations and introductions. These ancient texts—particularly the Mandukya, Chandogya, and Brihadaranyaka Upanishads—articulate the fundamental identity of individual consciousness (Atman) with universal consciousness (Brahman) that The Quantum Self recognizes as convergent with quantum insights.

Kashmir Shaivism

The Doctrine of Vibration: An Analysis of the Doctrines and Practices of Kashmir Shaivism

Mark S. G. Dyczkowski

The definitive scholarly introduction to Kashmir Shaivism, centering on the Spanda (vibration) doctrine. Dyczkowski explains how Shiva (pure consciousness) and Shakti (creative power) interact through primordial vibration to manifest reality—a framework that maps remarkably onto our discussion of how the QuantumLoom manifests as SpacetimeFabric. The book's treatment of the thirty-six tattvas illuminates the process of crystallization from unity to multiplicity.

Kashmir Shaivism: The Secret Supreme

Swami Lakshmanjoo

A distillation of Kashmir Shaivism's essence from the last living master of this tradition. Swami Lakshmanjoo presents

Abhinavagupta's profound synthesis in accessible terms, explaining consciousness as both the substance and the witness of all manifestation. His explanations of Shiva, Shakti, and Spanda provide the experiential context for the philosophical framework employed throughout The Quantum Self.

These ten sources represent starting points rather than comprehensive coverage. The convergence between quantum physics and contemplative traditions continues to generate new scholarship and insight. Readers are encouraged to follow the threads that resonate most deeply with their own inquiry.

NOTES ON THE NATURE OF CONSCIOUSNESS

In our pursuit to decode the nature of existence, to grasp the elusive threads of reality that weave the intricate tapestry of the cosmos, we inevitably arrive at the threshold of a profound enigma—consciousness. This mysterious phenomenon, the ineffable awareness that illuminates our perceptions, thoughts, and emotions, stands as the fundamental substrate of our understanding of reality.

For, what is reality, if not that which is sensed, perceived, and experienced through the lens of consciousness? What is a sunrise, if not the breathtaking splash of color witnessed at dawn, or a symphony, if not the harmonious melody that stirs the soul?

Within the labyrinth of our inquiry into the cosmos—its origin, its architecture, its very essence—we find consciousness subtly intertwined with every question, every hypothesis, every insight. Like an enigmatic sphinx guarding the gates of wisdom, consciousness challenges us to unravel its mystery before we can venture deeper into the secrets of existence. From the quarks to the quasars, from the neurons to the nebulae, every aspect of the universe is bathed in the light of consciousness, playing out a cosmic drama on the grand stage of awareness.

And so, at the end this book, we embark on a journey to explore this magnificent enigma. We navigate the intricate pathways of consciousness, traversing the terrains of philosophy, neuroscience, quantum mechanics, and age-old wisdom traditions. We dive deep into the depths of consciousness, seeking to illuminate the links that connect the observer with the observed, the perceiver with the perceived. It is a quest not just to understand reality, but to comprehend the lens through which we perceive it, the grand symphony of consciousness that orchestrates our experience of existence.

Consciousness: The Medical Perspective

Consciousness, to a medical professional, is often gauged with the term 'level of consciousness', a diagnostic tool used to ascertain a patient's responsiveness to stimuli. Consciousness here is the state of being awake, aware, and oriented. To a doctor, it's a life-saving thread that leads them to the mind of a patient, through the labyrinth of their physiology. Imagine, if you will, a doctor standing in a hospital room with a patient - a person rendered silent by unconsciousness, the typical language of words replaced by the language of biology.

The tools in a doctor's kit to decipher consciousness are simple yet profound. They ask, "Can the patient open their eyes? Can they respond to their name or obey commands? Can they differentiate between a spoon and a book?" Through this seemingly straightforward approach, the physician steps into the fascinating dance of neurobiology, assessing a patient's level of consciousness.

Physicians quantify consciousness using scales such as the Glasgow Coma Scale (GCS), a valuable tool born out of the realms of neurosurgery. The GCS judges consciousness on a numerical scale based on motor responses, verbal responses, and eye-opening. This scoring system bears witness to the power of medical objectivity when quantifying something as enigmatic as consciousness.

Yet, it is not only the level of consciousness that physicians discern. They strive to explore the quality of consciousness too. Delirium, a condition that marks the fluctuation in consciousness and cognitive impairment, is the medical epitome of altered quality of consciousness. It symbolizes the convergence of medicine and consciousness, leading doctors through a journey inside their patients' minds, using the compass of medical diagnosis.

Underpinning these assessments are the twin pillars of arousal and awareness - distinct, yet interwoven elements of consciousness. Arousal, governed by the intricate reticular activating system of the brainstem, controls wakefulness. In contrast, awareness is the canvas upon which the sensory inputs, emotions, and thoughts are painted, thanks largely to the interconnected corticothalamic network of the brain. The medical view of consciousness illuminates this deep entwinement of brain networks.

For physicians, understanding consciousness is akin to navigating the uncharted territories of a person's neural landscape, an endeavor where the topographical markers are neural networks, neurochemicals, and their interplay.

Consciousness: The Psychological Perspective

Our exploration of consciousness brings us to a different corridor in the labyrinthine castle of human understanding. Here, we unlock the door to the psychological perspective, a landscape where consciousness is viewed not merely as a neurological phenomenon, but also as an intricate tapestry of thoughts, emotions, desires, and memories.

In this realm, consciousness is often viewed as the sentinel of the human psyche, the watchtower that provides a panoramic view of our mental landscape. Here, consciousness is an ever-changing flow of mental experiences, a river carrying the flotsam and jetsam of our thoughts, emotions, and perceptions, creating the subjective narrative of our existence.

Psychologists often describe this narrative as a 'stream of consciousness,' a term coined by the American psychologist William James. This stream isn't a monotonous, unvarying flow; instead, it ebbs and surges, waxes and wanes, portraying the complex rhythms of our mental life. One moment, it might carry the bright pebbles of a joyful memory; the next, it might be clouded by the silt of worry or fear.

To navigate this ever-changing stream, psychologists often use the metaphor of a spotlight. The spotlight of attention, controlled by our consciousness, illuminates the specific thoughts, emotions, or perceptions we wish to focus on, pushing the rest into the shadows of the subconscious. Like a dedicated lighthouse keeper, consciousness guides the beam of our attention, helping us navigate the turbulent waters of our mental stream.

However, this spotlight doesn't merely illuminate our mental world; it also shapes it. According to psychological theories of consciousness, what we consciously attend to can profoundly influence our thoughts, feelings, and actions. For example, if our spotlight of attention consistently illuminates negative thoughts, it might color our entire stream of consciousness with shades of pessimism or anxiety.

But this psychological canvas of consciousness doesn't end here. It extends to more abstract realms, like our sense of self and our experience of free will. Consciousness, from a psychological perspective, isn't just about being aware of our thoughts or emotions; it's about being aware that we're aware, a concept often referred to as 'meta-awareness'. It is this ability to reflect on our thoughts and feelings that provides the foundation for our sense of self.

Finally, it is worth noting that the psychological view of consciousness is a dynamic one. Psychologists acknowledge that our understanding of this elusive phenomenon continues to evolve, influenced by ongoing scientific research and philosophical debates. It's a constantly shifting puzzle, its pieces reshaped by the tides of knowledge and understanding.

Consciousness: The Neuroscience Perspective

As we continue our exploration of consciousness, we find ourselves at the edge of a new frontier—the realm of neuroscience. This modern discipline has made considerable strides in unearthing the underpinnings of consciousness, drawing a fascinating, albeit complex, map of the relationship between our mind and brain. But like every map, this one too has its mysteries and uncharted territories.

Think of our brain as a grand concert hall, its neural pathways humming with the symphony of neuronal firing, much like a grand orchestra filling the air with harmonious melodies. Now, most conventional neuroscience perspectives would posit that the grandeur of this neuronal symphony—the spiking electrical potentials, the firing of synapses—is the very essence of consciousness. However, this is where we tread into contentious terrain.

Consider this analogy: Imagine asking Siri a question on your iPhone. The device processes the query, lighting up circuits and engaging processors, connecting on the cellular network to multiple remote servers to access information, and ultimately serving you an answer. The flurry of electron activity in the iPhone hardware is akin to the neuronal activity within our brains, lighting up areas like a busy city at night. But would we attribute Siri's coherent responses to the mere flow of electrons?

Just as it's the underlying software that drives the electron flow, leading Siri to comprehend and respond intelligibly, it's compelling to think of the brain activity we observe as the *result* of consciousness, rather than the *cause*. It is here that we encounter the 'hard problem' of consciousness, a term coined by philosopher David Chalmers. While neuroscience has been fairly successful in explaining the 'easy problem'—how specific brain states correlate with specific mental states—it stumbles when attempting to explain how or why these states should give rise to a subjective experience.

In this light, it's worth considering that consciousness might not be a product of any specific physical process in the brain, but rather, akin to the software in our iPhone analogy, it could be a complex orchestration of information, a non-localized phenomenon that weaves together our thoughts, sensations, and emotions into the unified field of experience we understand as being conscious.

This perspective doesn't dismiss the value of neuroscience, but rather reframes its role in our understanding of consciousness. It's not unlike appreciating the musical score as the blueprint of a symphony, while recognizing that it's the conductor's interpretation and the orchestra's performance that breathe life into the notes. The neural correlates of consciousness then become the footprints of the ephemeral, dancing spirit of consciousness, not its origin.

Locating Consciousness in Spacetime vs. Locating Brain Activity in Spacetime

As we continue our journey through the labyrinth of consciousness, we come upon a divergence in the path: the divergence between locating consciousness in spacetime and locating brain activity in spacetime. This juncture, laden with philosophical quandaries and scientific questions, invites us to peer into the enigmatic relationship between the tangible and the intangible—the seen and the unseen.

To begin, we look at the prospect of locating brain activity in spacetime. We reside in an era where advanced technologies like fMRI and PET scans provide us with a real-time view of our neural interplay. We bear witness to the riveting spectacle of neurons firing, correlating with complex phenomena such as thoughts, sensations, and emotions. Each synaptic connection, each neuron firing, exists as a tangible event in spacetime. They are the actors on the stage of our mental theater, and we can see them perform their roles vividly. But, can we equate this orchestration of neurons to the director behind the scenes—the consciousness?

Now let's journey down the other path, that of locating consciousness in spacetime. To illustrate the complexities here, let us use the same simple analogy we used earlier for your iPhone when you ask Siri a question. The electrons flowing through its circuitry—the hardware—is akin to the brain activity we see on a scanner. However, the response you get, the information processed and delivered—that's the work of software. The software, like consciousness, isn't confined to the physicality of the iPhone. It operates beyond, reaching out to the cloud, connecting to multiple servers across the globe to provide you with an answer. It's everywhere and yet nowhere, not confined to a single point in spacetime.

The consciousness conundrum is similar. We can map where cognitive processes take place in the brain, but where is consciousness—the awareness of being, the 'software' of our existence—located? Just as the software's functionality is spread between the iPhone and the far reaches of the internet, consciousness too, seems to resist being tied to specific points in spacetime.

This enigma prompts us to question and reevaluate the nature of what we seek. Is consciousness a byproduct of brain activity journeying through spacetime or is it more akin to a quantum 'Consciousness Field', omnipresent yet intangible?

Consciousness: The Advaita Vedanta Perspective

In the grand symphony of philosophical exploration, there are few schools of thought that have dedicated their entirety to the singular pursuit of comprehending consciousness and its relationship to reality. Among these, shining like a brilliant star against the backdrop of antiquity, is the philosophical school of Advaita Vedanta. Rooted deeply in the profound wisdom of ancient India, Advaita Vedanta offers an illuminating exploration into the enigma of consciousness, its indivisibility, and its inseparable connection to the fabric of reality.

Advaita Vedanta, with its focus resolutely fixed on consciousness, paints a picture of reality that is as elegant as it is profound. It posits that the universe, in all its vast and varied splendor, is not an assembly of separate entities, but a singular, unified field of consciousness. Advaita, meaning non-dual, proposes a view of reality where consciousness isn't just a facet of existence, but the very substrate on which the cosmos dances. It is this perspective, this fascinating interplay between consciousness and reality, that lends Advaita Vedanta its unique richness and depth.

As we embark on the exploration of Advaita Vedanta, we traverse the intricate landscape of consciousness as it unfurls itself in the wisdom of this ancient school of thought. We journey from the concept of individual consciousness to the universal consciousness, delving into the notions of the self, the universe, and the interplay between the two. This exploration invites us not only to understand the reality outside us but also to awaken to the reality within, to the luminescent beacon of consciousness that lights up our perception of the cosmos.

The Nature of Consciousness

In Advaita Vedanta, consciousness is regarded as the fundamental essence of reality, and it is inseparable from Brahman, the ultimate reality.

Consciousness is the underlying principle that illuminates all experiences and manifests as the sense of "I" or the self-awareness that is present in every individual. The nature of consciousness can be understood through the following aspects:

1. Self-luminous: Consciousness is self-luminous or self-revealing. It is the light that illuminates all experiences, objects, and phenomena. It does not depend on any external source for its existence or illumination, unlike physical objects that rely on external light sources to be seen.

2. Unchanging: Consciousness is eternal, unchanging, and beyond time, space, and causation. While objects, thoughts, and experiences are subject to change, consciousness remains constant and unaffected by these fluctuations. It is the unchanging background against which all experiences unfold.

3. Non-dual: Consciousness is the ultimate reality, and all phenomena, including the individual unchanging soul or Self (Atman), are manifestations of consciousness. There is no separation or duality between consciousness and the objects it illuminates. This non-duality is the core teaching of Advaita Vedanta, which asserts that the ultimate truth is the oneness of existence.

4. Subjective: Consciousness is the subject, the ultimate knower, and the experiencer of all objects and experiences. It is

distinct from the objects it illuminates and cannot be objectified or known in the same way that physical objects can be known.

5. Universal: While consciousness manifests as the individual self or Atman, its nature is universal and all-pervasive. The consciousness that illuminates one's experiences is the same consciousness that illuminates the experiences of all beings. This universal nature of consciousness reinforces the non-dual teaching of Advaita Vedanta.

To fathom this abstract concept, consider this metaphor. Think of the universe as an endless ocean, and each individual life as a wave. Though each wave may appear separate and distinct, rising and falling in its unique rhythm, it is, at its essence, nothing more than the ocean itself. It arises from the ocean, exists within the ocean, and eventually dissolves back into the ocean.

Similarly, in Advaita's view, our individual consciousness is not a separate entity, but an expression of the Universal Consciousness.

The concept of consciousness as the fundamental essence of reality is a major milestone in our journey towards understanding the mysteries of reality.

The Three States of Consciousness

Advaita Vedanta offers a framework for understanding human experience through the analysis of three states of consciousness: waking (jagrat), dreaming (svapna), and deep sleep (sushupti).

1. Waking state (jagrat): The waking state is characterized by the conscious experience of the external world through the senses, mind, and intellect. In this state, the individual self, or the sense of 'I' (jiva), identifies with the physical body, and experiences of the world appear to be real and separate from the self.

2. Dreaming state (svapna): In the dreaming state, the individual self experiences a separate internal world created by the mind, based on memories, desires, and subconscious impressions. In the dreaming state, the sense of 'I' continues. The dream world appears to be real to the dreamer, but upon waking, it is recognized as illusory.

3. Deep sleep state (sushupti): Deep sleep is the state of complete unconsciousness, where there are no thoughts, perceptions, or experiences. The sense of 'I' disappears, and the potential for future experiences remains in seed form.

Advaita Vedanta posits that the consciousness that illuminates all three states remains unchanged and is the unifying factor among them. The three states are superimposed upon consciousness.

Chit and Chidabhasa: Pure Consciousness and Reflected Consciousness

Advaita Vedanta distinguishes between two aspects of consciousness: Chit (pure consciousness) and Chidabhasa (reflected consciousness).

This distinction helps elucidate the relationship between the ultimate reality (Brahman) and the individual self (jiva) in the context of consciousness.

• Chit (Pure Consciousness): Chit is the original, unchanging, and infinite consciousness that is the essence of Brahman. It is the ultimate reality, free from all attributes and distinctions. Pure consciousness is self-luminous, self-existent, and the substratum of all creation. It is the one reality behind the apparent multiplicity of the world, and it transcends all limitations of time, space, and causation.

• Chidabhasa (Reflected Consciousness): Chidabhasa, or reflected consciousness, refers to the apparent individual consciousness that arises due to the reflection of pure consciousness on the subtle body, or 'inner persona', which is composed of the mind, intellect, and ego.

Chidabhasa is the limited, conditioned consciousness that identifies with the body, mind, and ego, creating the sense of a separate self or jiva. It is subject to change, being influenced by thoughts, emotions, and perceptions.

The analogy of the sun and its reflection on water is often used to explain the relationship between Chit and Chidabhasa. The sun represents pure consciousness (Chit), and the water represents the subtle body. The reflection of the sun on the water symbolizes the individual consciousness (Chidabhasa). The reflection appears to be a separate entity, but it is ultimately dependent on the sun and the water, just as the individual self is dependent on pure consciousness and the subtle body.

This concept of Pure and Reflected Consciousness, Chit and Chidabhasa, finds strong parallels in quantum mechanics, as we saw in Chapter 2.

We Are The Universe

Our venture into the realms of Advaita Vedanta brings us to a fascinating proposition - that of the universe as a projection of consciousness, a play of cosmic proportions projected upon the grand screen of Brahman, the ultimate reality. This concept, inherently abstract and metaphysical, finds unexpected parallels in modern quantum physics, particularly in the 'observer-dependent' view of reality.

In the cosmic dance of Advaita, Brahman—pure, unbounded consciousness—is the singular reality, the foundational substrate upon which the universe is superimposed. There is no duality, no multiplicity of existences separate from Brahman. The stars in the night sky, the roaring ocean, the mountains standing sentinel—all are expressions of Brahman's consciousness, projected onto the canvas of existence.

Interestingly, this idea mirrors the Copenhagen interpretation of quantum mechanics where the act of observation influences the observed. A particle exists in all its theoretically possible states simultaneously until it is observed. It is the act of measurement—or observation—that 'collapses' the wavefunction into one possibility.

Current quantum mechanics interpretations like the Copenhagen interpretation posit that consciousness—specifically, the observer's consciousness, or Chidabasa, to use the term from the previous section—becomes an inextricable part of the process that defines reality.

We explored this in Chapter 2, to see if it is Chit – Pure Consciousness, that causes the collapse of the wavefunction, and not the observers consciousness or Chidabasa.

Consciousness Determines Reality

In the vast sea of possibilities that quantum mechanics allows, it is the act of conscious observation that pinpoints a single reality from the multitude. The question then arises: Is it possible that the entire universe exists in a state of superposition, with every possible event happening at once, until observed by conscious entities? This perspective raises intriguing possibilities, a meeting point of Eastern philosophy and quantum physics, both postulating that consciousness plays a pivotal role in determining reality.

Advaita Vedanta's view further extends, proposing that the observer, the act of observation, and the observed are not separate, but all are expressions of Brahman. This non-dual perception aligns remarkably with physicist John Wheeler's "Participatory

Anthropic Principle," which suggests that we are participators in bringing about the universe just by our act of observation.

Here, consciousness is not merely an isolated phenomenon; it's the substratum of existence, the universal principle interlacing matter and mind. We find ourselves reconsidering the very nature of reality, causality, and the complex interplay between observer and observed, catapulted into a realm where consciousness reigns supreme.

Consciousness Is Not Within Us, We Are In Consciousness

In the realm of Advaita Vedanta, consciousness is no small, confined entity, no mere function of the physical brain ensnared within the limits of space and time. Instead, it is perceived as an expansive, boundless field—infinitely vast, pervading all of existence.

To the Advaitic seer, consciousness is not merely housed within us, but we are, in fact, housed within Consciousness. Our sense of self emerges not from our biological entity but arises from this undivided, all-encompassing sea of Consciousness. Like a wave arising from the ocean, our individuality is nothing but an expression of this grand, underlying field of conscious awareness.

This vision is often compared to the relationship of space contained within a pot, known as "ghatakasha", with the limitless sky or "mahakasha". When the pot is formed, it seems to enclose a portion of the vast sky. However, in reality, the space within the pot is not separate from the limitless sky outside. The partition is merely apparent, and the seeming inside-outside dichotomy is simply a result of our limited perspective.

Similarly, we might feel that consciousness is confined within us—within our minds, our bodies. But this is akin to the illusion of the pot-space—our limited perspective veils the greater truth that our individual consciousness is not separate but an indivisible part of the universal consciousness.

The recognition of this profound truth—that we exist within the field of consciousness rather than consciousness existing within us—lies at the very heart of Advaita Vedanta. It dismantles our conventional understanding of self and universe, leading us towards the recognition of a greater, non-dual reality. Here, there is no 'inside' or 'outside', no 'me' or 'other', but one seamless, interconnected expanse of conscious awareness.

Kashmir Shaivism and Consciousness

In the elegant philosophies that have emerged from the Indian subcontinent, few are as rivetingly mystic and profoundly transformative as the tradition of Kashmir Shaivism. An esoteric spiritual philosophy emanating from the snow-clad peaks and verdant valleys of ancient Kashmir, it proposes a breathtaking view of consciousness. Unlike many Western philosophies, which often side-line consciousness as a byproduct of matter, Kashmir Shaivism places it at the epicenter of existence, perceiving it as the pulsating heart of all reality.

In Kashmir Shaivism, consciousness, termed as "Shiva," is not confined to the limited perspective of the individual. Instead, it unfolds as an infinite, universal continuum—a dynamic, all-embracing field that encapsulates both the observer and the observed, the experiencer and the experienced. This sweeping canvas of consciousness, in its ceaseless dance of creation, preservation, and dissolution, manifests the universe in its limitless diversity. Here, consciousness is not a passive spectator but an active participant in the grand theater of existence, playing out every role with exquisite perfection.

This universal consciousness, far from being an abstract, remote concept, is the very essence of our being—the intimate, immediate sense of awareness that pervades every moment of our experience. It is both the silent witness and the dynamic player in the dance of life.

The study of consciousness in Kashmir Shaivism is akin to diving deep into this symphony, not just as a listener, but as a note, a rhythm, an integral part of the melody, resounding in the infinite expanse of the conscious universe.

As we voyage deeper into the realm of Kashmir Shaivism and its understanding of consciousness, we encounter the fascinating concept of the 36 Tattvas. The term 'Tattva' is Sanskrit for 'truth' or 'reality.' It represents the stages or principles through which the Absolute Consciousness, Shiva, manifests into the tangible, physical universe we perceive.

This gradual unfolding of consciousness, like a cosmic drama that traverses the realm of the absolute to the realm of the relative, is meticulously mapped in the 36 Tattvas.

The first five Tattvas, often termed as the 'Pure Tattvas,' concern the realm of the Absolute, where duality does not exist. The first Tattva, Shiva Tattva, represents the undifferentiated Absolute Consciousness, which is pure, infinite, and unbounded.

As we move to the next Tattvas - Shakti, Sadasiva, Ishvara, and Sadvidya - we witness a subtle transition where Consciousness, though remaining absolute, begins to entertain the possibility of creation, the notion of 'otherness,' and a universe to manifest.

From the 6th to 11th Tattvas, termed 'Pure-Impure Tattvas,' we enter a phase where the manifestation process becomes more pronounced. They encompass the five fundamental aspects of cosmic existence: creation, sustenance, dissolution, concealment, and grace. Each of these stages mirrors a significant shift in consciousness, symbolizing a move towards individualization and manifestation.

The final 25 Tattvas, the 'Impure Tattvas,' comprise the perceptible universe and its constituents. Here, consciousness dons the cloak of duality, manifesting as individual souls, the mind, the senses, and the elements of nature. The 36th Tattva, Prithvi, symbolizes the physical earth, the final stage of Consciousness's journey from formless unity to tangible diversity.

This procession of Tattvas is not a descent into lower states but rather an exploration of consciousness's infinite possibilities.

In the grand theater of existence, we, as individual beings, are not separate from the Absolute Consciousness but are its dynamic expressions. Our individual consciousness, ensconced in our human experience, is a microcosmic echo of the macrocosmic Consciousness.

Kashmir Shaivism, through the metaphor of the Tattvas, allows us to appreciate the vastness of Consciousness and our profound interconnectedness within it. This understanding, as we progress, promises to add another dimension to our exploration of the relationship between consciousness and the universe.

You Are A Reflection Of Supreme Consciousness

Imagine standing in front of a mirror. You see your reflection, recognize it as 'you,' but you also understand that the reflection is not 'you' in essence. It is merely a representation of you. Now, apply this analogy to your perception of the universe.

Pratyabhijna, in Kashmir Shaivism, invites us to perceive the universe, not as separate, external entities, but as a reflection of the Supreme Consciousness, which is our true nature.

This Recognition isn't about acquiring new knowledge or attaining a higher state. Rather, it's about realizing what has always been true. It's about recognizing that the individual self or 'I-consciousness,' immersed in the world of sensory experiences, is not distinct from the Supreme Consciousness.

Kashmir Shaivism proposes that this recognition isn't an intellectual or conceptual understanding, but a profound, direct realization. It suggests a shift in our perception, where we begin to 'see' the world, not through the prism of duality but through the lens of non-duality. We realize that the world of multiplicity is, in truth, a dance of the singular Consciousness.

Pratyabhijna represents an 'aha moment' in our consciousness journey, a moment of clarity when the boundaries of individuality blur into the vastness of universality. In this state, there's no 'me' or 'you,' no 'observer' or 'observed.' All there is, is Consciousness.

Through the lens of Pratyabhijna, our exploration of consciousness broadens and deepens. We begin to appreciate that consciousness isn't just an abstract, impersonal concept but an intimate, lived reality and invites us to revisit our understanding of ourselves and our place in the universe.

Spanda - The Pulsation of Consciousness

The story of Consciousness in the framework of Kashmir Shaivism is not one of a static, unchanging entity, locked away in some ethereal realm. Instead, Consciousness, or Shiva, is described as dynamic, vibrant, pulsating. This idea is encapsulated in the concept of 'Spanda,' a Sanskrit term roughly translating to 'pulsation,' 'vibration,' or 'throbbing from within.' Spanda isn't just a characteristic of Consciousness; it is Consciousness itself, as seen through the lens of its unceasing activity.

Think of a quiet, serene pool of water. At first glance, the water appears calm, almost stagnant. But as you draw closer, you notice ripples forming spontaneously on its surface. No external force seems to be causing these ripples; they seem to be born from the very heart of the water. This is Spanda – the inherent, self-sustained pulsation of Consciousness.

Kashmir Shaivism postulates that it's through Spanda that Shiva, the Supreme Consciousness, projects Himself into the universe. This projection isn't a departure from the source; rather, it's an expansion, an overflow of Consciousness's vibrancy. It's through this dynamic, throbbing energy that the unmanifest becomes manifest, the singular becomes multiple, and Consciousness objectifies itself in the form of the universe.

The doctrine of Spanda bridges the gap between the unmanifest Shiva and the manifest universe. It provides a mechanism, so to speak, for how the Absolute can manifest as the relative, how the infinite can appear as the finite, and how the unbounded can take on boundaries. It offers a window into the paradoxical play of Consciousness, as it veils and unveils itself in the cosmic drama.

Spanda And Quantum Physics

This principle, intrinsic to the philosophy's heart, resonates with intriguing parallels to the enigmatic realms of quantum mechanics and string theory.

To the quantum physicist in the realms of quantum field theory, the fabric of reality oscillates with ceaseless fluctuations. It's an intricate dance of particles and antiparticles, popping in and out of existence in the fleeting time frames permitted by Heisenberg's uncertainty principle, in what has come to be known as the quantum vacuum or vacuum energy. In this microcosmic dance hall, there is no such thing as 'empty' space. The so-called 'void' pulsates with potentiality, reverberating with the infinite possibilities of creation.

Echoing this concept, Spanda in Kashmir Shaivism represents the divine creative pulsation—the vibratory dynamism through which the supreme consciousness manifests the universe.

At the forefront of modern physics, string theory proposes that the basic constituents of reality are not point-like particles but one-dimensional 'strings'. These strings vibrate at different frequencies, each frequency corresponding to a different particle type. The entire universe, in this view, is a symphony of vibrating strings. Drawing parallels to Spanda, the universe, as Kashmir Shaivism suggests, is also a symphony, a vibratory play of consciousness—a manifestation of the divine pulsation.

From the rhythms of Spanda to the enigmatic vibrations of quantum mechanics and string theory, we uncover a sublime meeting point—a space where ancient wisdom converges with modern scientific insight, each enriching the other's narrative.

The captivating dance of Spanda reverberates not just within the heart of the seeker, but also within the heart of the cosmos, reflecting the profound dialogue between science and spirituality that this exploration of consciousness endeavors to capture.

Quantum Consciousness Theories

The cradle of quantum consciousness is nestled in the paradoxes and peculiarities of the quantum realm. Think about the wave-particle duality, the observer effect, and the collapse of the wave function. All of these phenomena hint at a reality that is far removed from our everyday experiences—a reality that seems intertwined with consciousness. Quantum consciousness theories arise from a desire to bridge this apparent chasm between the material and the mental, between the quantum and the conscious.

One of the most prominent theories in this arena is the Orchestrated Objective Reduction (Orch-OR) theory, championed by the likes of

Sir Roger Penrose (2020 Nobel Prize for Physics) and Dr. Stuart Hameroff. They propose that consciousness arises from quantum computations occurring within the brain's microtubules. According to them, the collapse of quantum superpositions within these microtubules generates conscious awareness, like a beautiful symphony arising from a well-orchestrated interplay of individual notes.

A recent groundbreaking study has provided experimental evidence suggesting a quantum basis for consciousness. By demonstrating that drugs affecting microtubules within neurons delay the onset of unconsciousness caused by anesthetic gases, the study supports the quantum model over traditional classical physics theories.

But what implications does this have for our understanding of consciousness? If consciousness indeed arises from quantum processes, it means our understanding of reality—and our place within it—might need a radical reevaluation. No longer would consciousness be confined to the realm of the ethereal or the subjective. Instead, it would take its rightful place alongside other fundamental aspects of reality, such as matter and energy.

These perspectives open up an exciting new frontier in our quest to understand consciousness. They urge us to rethink, reimagine, and reassess. They stir up the waters of scientific discourse, inviting new ideas, fostering rich debates, and nudging us closer to that elusive understanding of consciousness.

In the end, whether quantum consciousness theories hold up under the scrutiny of scientific examination or not, they have already succeeded in a significant way. They have ignited a conversation, a vital dialogue that is helping to propel our understanding of consciousness forward. They remind us that consciousness, far from being a peripheral curiosity, lies at the very heart of our quest to understand the universe—and ourselves within it.

Summary: The Relationship Between Consciousness And Reality

Let us now summarize the three fundamental pillars on which our understanding of the relationship between Consciousness and Reality stands, at this point in our journey.

The First Pillar: Consciousness and Awareness Are The Only Known Realities

From the hallowed halls of academia to the most private corners of our minds, the exploration of consciousness has been at the forefront of human inquiry.

Our search for understanding has taken us across a vast intellectual landscape, encompassing myriad perspectives, theories, and philosophies. But through all these contemplations, one fact rings out, resonating with the clarity of a temple bell: Consciousness, and its cousin, awareness, are the only known realities to us.

Embark upon a thought experiment. Consider your life, the stream of experiences you have had since you were born. What is the constant thread binding all these diverse moments, the one enduring fact? The answer is your consciousness, the ever-present 'I' that is aware of all these experiences. It's as simple as it is profound: we are conscious beings. Whether we are savoring the tang of a ripe apple, immersed in the hypnotic rhythm of rain on a rooftop, or ensnared by the haunting melody of a symphony, it is consciousness that breathes life into these experiences.

Each one of us lives within our own conscious world, a world populated by perceptions, thoughts, and emotions. And all these elements of our world are known to us through awareness. Every sight we see, every sound we hear, every texture we touch, every scent we inhale, every taste we savor, every emotion we feel, every thought we think – none of them exist as independent objects , they all arise within our consciousness and are perceived by our awareness.

Consciousness and Awareness Are Two Different Things

In our ongoing journey to unravel the intricacies of consciousness, we now stumble upon another twist in the plot – the distinction between consciousness and awareness.

This differentiation, subtle but significant, serves as the bedrock of our understanding of our conscious experiences and our awareness of these experiences. While often used interchangeably in colloquial language, in the nuanced exploration of the mind and the nature of reality, they carry distinct meanings.

Firstly, let's grapple with the idea of consciousness. As we have discussed previously, consciousness is the individual, personal, subjective experience that each one of us possesses. It's a vast, internal universe unique to each individual, brimming with thoughts, perceptions, emotions, memories, and dreams.

However, consciousness does not exist in isolation. It needs an observer, a presence that perceives these thoughts and sensations. This presence is what we refer to as 'awareness'. Awareness is the clear, open space in which the contents of consciousness manifest themselves. If consciousness is the play, then awareness is the stage upon which the play is performed. It is the silent, ever-watchful observer, undisturbed and unaffected by the events taking place within consciousness.

To use an analogy, think of consciousness as the images being displayed on a movie screen, while awareness is the screen itself. The images – our thoughts, emotions, and perceptions – can be vivid, intense, or subtle, filled with various colors and shapes, constantly moving and changing.

However, throughout all this activity, the screen – our awareness – on which the images are displayed, remains unchanged. It allows the images to appear, to play out their roles, and to fade away, but it itself remains untouched and constant.

This screen is not bound by the images it displays. The same screen that displays a story set in ancient times can effortlessly transition to showcase a narrative set in a distant future, effectively transcending the limitations of time. It is also not restricted to a specific location in space - regardless of whether the story unfolds in the deepest oceans or the farthest reaches of space, the screen remains. Thus, awareness is essentially outside time and space.

The Witness

Advaita Vedanta uses the concept of the 'Witness' (Sakshi) as a means to recognize the Atman - the true Self that is pure consciousness itself.

This 'Witness' is our pure awareness, the silent observer of our internal universe. It is distinct from our thoughts, emotions, and perceptions, yet it is through this 'Witness' that we perceive them. It stands apart, watching without judgment, without preference, and without any involvement. It's the same awareness that observes our thoughts, that perceives a beautiful sunset, that experiences the taste of food. It is the common thread running through all our experiences.

Awareness is pure, unadulterated, and ever-present, even in the absence of conscious thoughts or perceptions. This intricate dance between consciousness and awareness, between the watched and the watcher, between the observed and the observer, forms the core of our understanding of the self and the nature of reality. This awareness is not just what gives rise to our sense of 'I,' but is our true Self beyond all limited identifications.

The practice of witnessing serves as a stepping stone to this recognition. By identifying with the unchanging awareness rather than the changing content of experience, we begin to recognize our true nature. But in the highest understanding of Advaita, even the witness-witnessed duality dissolves into pure, non-dual awareness.

In this ultimate perspective, there is no separate consciousness containing experiences and no independent awareness observing them. There is only Brahman—pure consciousness that is both the knower and the known, the seer and the seen, the one reality appearing as the multiplicity of experience while remaining forever one and undivided

Reality Does Not Exist Without Consciousness

Imagine a world without consciousness. What would it be like? It wouldn't 'be' anything, for there would be no awareness to perceive it. A universe without consciousness to perceive it is akin to an unopened book—an aggregation of information with no one to read and understand it. Hence, our conscious awareness is what lights up the world, making reality a reality.

From the medical, psychological, and neuroscientific perspectives to the profound insights offered by Advaita Vedanta and Kashmir Shaivism, the prominence of consciousness is universally acknowledged. Even quantum mechanics, with its counterintuitive principles and bizarre implications, seems to be pointing towards a universe where consciousness plays an integral role.

In essence, consciousness is the canvas upon which reality is painted, and awareness is the light that illuminates this canvas. They are the silent witnesses to the drama of existence, the anchors that hold the fabric of our reality together. This understanding of consciousness and awareness, as the only known realities, is not just an intellectual proposition—it is a lived experience, the very bedrock of our existence.

The Second Pillar: Consciousness Is Beyond The Confines Of Space And Time

Like the subtle notes of a symphony carried across the stillness of a concert hall, our exploration of consciousness echoes through the realms of time and space. However, as we continue our intellectual journey, we arrive at an intriguing junction—a crossroads where our everyday perceptions of reality are thrown into question. Here, we are asked to consider the nature of consciousness itself, not as it relates to the physical world, but in its own right. And in this light, we begin to see consciousness not confined within the boundaries of spacetime but transcending it—non-spatial and non-temporal.

Let's contemplate a bit further. Where, for example, does the experience of savoring a piece of dark chocolate happen? Where does the exhilaration of an insightful idea originate? These questions may seem a tad unusual at first, but they reveal a profound truth. The experiences do not occur in the chocolate nor in the neural activity within our brain. They emerge in our consciousness, a domain unbound by physical coordinates of spacetime.

Similar to the concept of zero in mathematics, which is neither positive nor negative and yet a vital foundation for our numeric system, consciousness is beyond the duality of space and time. It isn't 'located' anywhere and yet it 'houses' all experiences. Try as we might, we cannot pin consciousness to a specific point in time or a particular place in space.

Even in the realm of neuroscience, this realization holds true. Brain activity—the firing of neurons, the electric currents, the chemical reactions—can be measured and mapped within the dimensions of spacetime. However, the consciousness that perceives these signals, the 'I' that experiences the perception, remains elusive to spatial or temporal coordinates.

Our understanding of consciousness, as gleaned from the wisdom of the Vedas, reinforces this non-spatial, non-temporal nature of consciousness. The Vedas portray Brahman, the ultimate reality akin to our concept of consciousness, as 'Neti, Neti'—'not this, not this.' This implies that consciousness, much like Brahman, cannot be confined to any specific phenomenon or construct within the dimensions of spacetime.

Similarly, Kashmir Shaivism, with its nuanced explanation of the 36 tattvas, illustrates the intricate unfolding of the universe from the unmanifest Shiva Consciousness. But the source, the Shiva Consciousness itself, is beyond the constraints of spacetime. It's the eternal stage upon which the cosmic drama unfolds, yet it is not confined within the duration of the play or the dimensions of the stage.

In the realm of quantum mechanics, we glimpse echoes of this understanding. The wave function, representing the state of a quantum system, exists in a superposition of states until measured or observed. The act of observation 'collapses' this wave function into a specific state. The observer effect, as it's known, implies that the act of observation, an act of consciousness, lies outside the conventional spacetime framework.

So, whether we're examining the nature of consciousness from the perspectives of neuroscience, spirituality, or quantum physics, we find it extending beyond the dimensions of space and time. This non-spatial, non-temporal characteristic of consciousness challenges our standard perceptions, inviting us to contemplate reality beyond our conventional understanding.

Thus, it seems consciousness is like a timeless, boundless ocean, with the waves of thoughts, perceptions, and experiences arising and subsiding within it. Regardless of where these waves emerge or when they dissolve, the ocean remains—an infinite expanse beyond the constraints of spacetime, cradling within it the possibilities of all existence.

The Third Pillar: Distinguishing Between Consciousness And consciousness

As we continue our exploration of consciousness, we arrive at a point where we encounter a peculiar nuance, a subtle but important distinction that has profound implications for our understanding of this enigmatic phenomenon.

This distinction lies in the use of a single letter - the letter 'C'. It is the difference between Consciousness with a capital 'C' and consciousness with a lowercase 'c'. This differentiation, while seemingly minimal, captures a fundamental concept in our quest to unravel the mysteries of consciousness.

Let's start with 'consciousness' with a lowercase 'c'. When we refer to consciousness in this sense, we are referring to the individual awareness of one's thoughts, feelings, and perceptions. It is the private, first-person experience unique to each one of us. It is your personal awareness of the color blue or the taste of an apple, the sensation of sunshine on your face, or the sound of a song that stirs deep emotions within you. This consciousness, subjective and deeply personal, is what grants us our individuality, our sense of self.

However, as we step back and consider the wider picture, we encounter Consciousness with a capital 'C'. This Consciousness refers to a universal, all-encompassing field of awareness, akin to what the Advaita Vedanta tradition refers to as Brahman or what Kashmir Shaivism describes as Shiva Consciousness. It is a fundamental aspect of reality, unbounded by the personal experiences or subjective perceptions of individuals.

This Consciousness is a timeless, spaceless continuum of awareness that permeates all existence. It is the ocean within which the waves of individual consciousnesses arise and subside. It is not confined to the perceptions of a single observer, nor is it limited by the boundaries of spacetime. It is, in essence, the source from which all individual consciousnesses spring and to which they return.

The dance between Consciousness and consciousness, then, becomes a dance between the universal and the individual, the absolute and the relative, the ocean and the wave. It's a cosmic ballet that plays out across the stage of existence, weaving a tapestry of experiences that contribute to our understanding of reality.

Understanding this dance, and the distinction between Consciousness and consciousness, brings us one step closer to understanding our place in the cosmos. It offers a profound insight into the nature of reality and our role within it. For it implies that while we each experience the world in our unique ways, these experiences arise within, and are part of, a grand, all-encompassing Consciousness.

AUTHORS REFLECTIONS

As a nine-year-old boy, I lived with my grandparents in the town of Coonoor, located in the Nilgiris (Blue Mountain) range in southern India. A morning that is forever etched in my memory. Our house stood atop a hill, overlooking a landscape of rolling hills and a river that wound its way through the valleys. The air was crisp and cold, and a dense fog had settled over the land, obscuring the world below in an ethereal haze.

I found myself lying in the grass, gazing down at the fog-shrouded landscape. The mist seemed to have a life of its own, rolling up the hill in gentle waves, alternately concealing and revealing the secrets of the terrain. It was as if the fog was playing a game of hide-and-seek with reality, teasing my imagination with fleeting glimpses of what lay beneath.

As I watched, the sun suddenly broke through the haze, its golden rays piercing the veil of mist. In that moment of perfect clarity, the world below me was illuminated in stunning detail. The fog, though still present, could no longer obscure the truth that had been hidden from view. It was as if a curtain had been drawn back, exposing a deeper reality that had always been there, waiting to be discovered.

In that instant of revelation, I felt an overwhelming sense of connection to the universe. Every fiber of my being seemed to be interwoven with the fabric of existence itself, and I experienced a profound sense of oneness with a universal consciousness. It was as if I had tapped into a vast, infinite intelligence that permeated all things, binding them together in a tapestry of meaning and purpose.

As I lay there, bathed in the warmth of the sun and the glow of this newfound understanding, an intense curiosity and a deep desire to explore the meaning of it all welled up within me. I knew, with a certainty that defied my young age, that this moment had set me on a path that would define the course of my life.

The seeds of curiosity and the quest for truth, planted on that misty morning in Coonoor, continued to flourish throughout my life, nurtured by the unique blend of influences that shaped my upbringing.

I would like to share this journey with you, the paths I took and how I came to write this book.

My family was a tapestry woven from threads of both the practical and philosophical worlds, each contributing to the richness of my intellectual and spiritual journey.

My paternal grandfather, a distinguished judge, embodied the principles of logic, reason, and the rule of law. His presence in my life instilled in me a deep respect for the power of the rational mind and the importance of critical thinking. On the other hand, my maternal grandfather, a lawyer by profession, had a keen interest in the ancient wisdom of Indian philosophy. Through him, I was introduced to the profound insights and timeless truths that lay hidden within the pages of ancient texts.

My father, a brilliant physicist, had been pursuing his Ph.D. in Physics at the Imperial College in London when the outbreak of World War Two compelled him to return to India. Despite this unexpected change in his path, he adapted and thrived, becoming an executive in a leading company. His journey taught me the value of resilience, adaptability, and the ability to find success no matter where the path may lead.

Yet, amidst the practicality and worldly pursuits of my family, there was a strong undercurrent of philosophical inquiry and a deep reverence for philosophical traditions. My mother, a qualified teacher of Sanskrit, the ancient language in which the great Indian philosophical treatises were written, opened the doors to a world of profound wisdom and spiritual insight. Her teachings and the exposure to these ancient texts left an indelible mark on my young mind, shaping my perception of reality and igniting a lifelong quest for understanding.

As I grew older, I found myself drawn to the study of physics, captivated by the elegant laws and principles that governed the universe. After graduating with a degree in this field, I completed my master's in business management and then embarked on a journey that would take me through the realms of business and technology, each step a reflection of my innate curiosity and the desire to explore new frontiers.

I began my professional life in a large conglomerate in India, but soon found myself restless, yearning for a new challenge that would push me beyond the boundaries of the familiar. This led me to join a much smaller company making cooking oils, and I took them into the Information Technology business as the Founding CEO.

I had the opportunity to help create and shape this venture in the information technology field that was, at the time, still in its nascent stages. That company, now public on the New York Stock Exchange, has grown to become one of the largest players in its domain.

My personal journey then took an unexpected turn when my daughter's health required specialized medical treatment, prompting a move to Silicon Valley in the San Francisco Bay Area. In this new environment, I found myself drawn to the dynamic world of startups, where the spirit of innovation and the desire to create something groundbreaking pulsed through the very air.

I joined a small startup company and played a pivotal role in its global growth, witnessing firsthand the exhilaration of taking an idea from conception to global fruition. After the company's successful IPO, I embarked on a new chapter, becoming the Founder and CEO of four venture-capital-funded technology startups, some of which became billion dollar corporations. These ventures spanned diverse domains, from payments and e-commerce to artificial intelligence, each presenting its own unique set of challenges and opportunities.

Looking back, I can see a common thread that ties together these seemingly disparate experiences: curiosity and a constant search for new challenges. Just as I had been drawn to the cracks and hidden truths in the misty landscape of Coonoor, I found myself seeking out the unexplored spaces in markets and products, the gaps where innovation could take root and flourish.

Throughout this journey, I never lost sight of my passion for physics. Even as I navigated the fast-paced world of business and technology, I continued to learn and stay abreast of the remarkable discoveries and advancements in the field. This ongoing engagement with physics not only fed my intellectual curiosity but also provided a foundation for the ideas that would later take shape in this book.

Over the last decade, I have had the privilege of collaborating with my friend Menas Kafatos Ph.D., an Endowed Professor of Physics at a leading university. Together, we have published several technical papers in leading physics journals probing the intricacies of the universe and exploring the frontiers of our understanding. Our shared passion for physics, consciousness, Greek and Indian philosophy and our willingness to question the boundaries of conventional thinking sparked a journey of intellectual exploration that has been both exhilarating and transformative.

But it was perhaps my exploration of ancient philosophical traditions from India that completed the picture in unexpected ways. As I delved into the non-dualistic teachings of Advaita Vedanta and the vibrational cosmology of Kashmir Shaivism, I was struck by the resonances between these millennia-old wisdom traditions and the cutting-edge discoveries of modern physics.

In the Advaitic concept of Brahman—the singular consciousness underlying all existence—I found echoes of the quantum field that pervades all space. In Kashmir Shaivism's dance of Shiva and Shakti, I recognized the interplay between potentiality and manifestation that quantum mechanics describes through wavefunctions and observation.

These traditions weren't just poetic metaphors or primitive attempts to understand the universe—they were sophisticated frameworks that had intuited, through deep contemplation rather than experimental science, some of the same fundamental truths we're now rediscovering through quantum theory. The universe they described—unified, consciousness-linked, vibrational at its core—aligned remarkably with the picture emerging from our most advanced physics.

This realization was not merely intellectual but transformative, compelling me to seek a framework that could bridge these worlds: the quantifiable precision of physics and the experiential wisdom of ancient contemplative traditions.

The Unified Conscious Matrix theory presented in this book is the fruit of this synthesis—not a final answer by any means (I do not believe that such a thing exists), but an invitation to see reality afresh, to recognize that the apparent division between the quantum and classical worlds, between mind and matter, between observer and observed, may be an artifact of our limited perspective rather than a fundamental truth of existence.

This book is my attempt to share this journey of discovery, to invite you to glimpse the universe not as a collection of separate objects and forces, but as a magnificent whole—a cosmic tapestry where every thread, including our own consciousness, is intrinsically woven into the grand pattern of all that is.